Twenty-First Century
ADVANCED
CHEMISTRY

HARMINDER GILL

authorHOUSE®

AuthorHouse™
1663 Liberty Drive
Bloomington, IN 47403
www.authorhouse.com
Phone: 1 (800) 839-8640

Published by AuthorHouse 05/05/2015

ISBN: 978-1-5049-0594-7 (sc)
ISBN: 978-1-5049-0617-3 (e)

Print information available on the last page.

Contents

Acknowledgements

I wish to acknowledge Author House for taking the time to publish this book. Thank you very much for the time and dedication to make this book possible.

Preface to the General Reader

How do we define chemistry? Multiple authors have defined chemistry in different ways. Chemistry can be thought of as the science of change at the microscopic level. Change is always taking place. This includes the planets of our solar system, substances being formed and decayed from matter, or the simple tasks we carry out in life such as walking and breathing. We could not live comfortably without the changes our planet goes through. In fact, the changes that take place are continuously published in journal articles and in new books. Chemistry is a broad field that encompasses many different fields. It is impossible to keep current in every single field. There are five main different fields in chemistry. Organic chemistry is the study of the element carbon and its reactions. Inorganic chemistry is the study of the rest of the elements on the periodic table. Analytical Chemistry deals with the analysis of chemical compounds. Physical chemistry deals with the physical properties of substances and their composition. Finally, biological chemistry deals with the chemistry that takes place in the biological world. There are other fields in chemistry which is atmospheric chemistry, environmental chemistry, engineering chemistry, the chemistry of materials, medicinal chemistry, and many more fields which integrates topics covered in chemistry.

This book was made for the advanced reader who already has a strong background of general chemistry, analytical chemistry, and organic chemistry. The topics covered in general chemistry are broad. You will find some of the topics in this book to be almost the same and some of the topics will be a bit different. Some of the later chapters includes an analysis of organic compounds, the chemistry of our planet Earth, the periodic table of elements with more elements that have been synthesized, the chemistry of art, and cosmetic chemistry. The material presented is not stagnant. Research is always being carried out throughout the world, and could replace some of the content presented in this book in the future.

Regardless of the chemistry course you took or if you have never seen topics covered in chemistry, you will find that the microscopic world of chemistry is different from the macroscopic world in which we live in. Both worlds are highly connected to each other. The chemistry that takes place in the microscopic world greatly affects the chemistry that takes place in the macroscopic world. Currently, we are in the twenty-first century, and at this day and age we have a stronger understanding of atoms, molecules, compounds, and reactions than we did almost one hundred years ago due to upgraded technology. But we still have many more questions that remain unanswered such as how and why some reactions take place and an accurate representation of most of the mechanisms that place in reactions.

Chemistry requires a strong foundation of algebra and calculus even for a graduate level course in organic and inorganic chemistry. Both spectroscopy and group theory deal with equations of molecules. Without a strong background in mathematics, it is almost impossible to understand the subject matter not just for chemistry but in many other scientific disciplines. Relationships such as one variable increasing or decreasing in comparison with another variable allows the scientist to formulate hypothesis, carry out experiments, and propose theories that can be tested over time. Theories can be accepted or rejected as new information is presented. Theories can become natural laws if they considered correct. A whole new world of chemistry still awaits.

I prepared this book to be as readable as possible and to be used as a reference if needed. Lots of material is covered in order to have a strong understanding of both the macroscopic and microscopic world. Understanding qualitative concepts is important before doing the quantitative problems. Determining the correct answer without having an understanding is a poor method to learn and a waste of time for instructors to teach. Memorizing the answers does not help when learning the material. Probing your understanding of the material as well as leading to the correct method and obtaining the correct answer is a better method to learn and master the material. You can learn a lot if you can explain the material to yourself. This gives you a method to test if you know the material and recognize any lack of understanding on your part.

Problem solving is a key concept you should get out of science courses. Whether you decide to major in chemistry or the some other science, you can still apply the methods of problem solving to their everyday lives. The order of topics is similar to courses taught at the college and university level. Some courses have there teaching topics in correlation with the current day's laboratory experiment. Others have their topics based on a teaching in preparation for standardized tests or a combination of the above. Each chemistry class is taught differently, but the material and the content remains almost the same depending what is covered and what is not covered. It is my hope that twenty-first century advanced chemistry will be a valuable book for all who want to learn the material at a higher level, and as an incentive to continue to learn more chemistry that takes place in our daily lives to solve the problems we face as a society.

Chapter 1
Overview of Chemistry

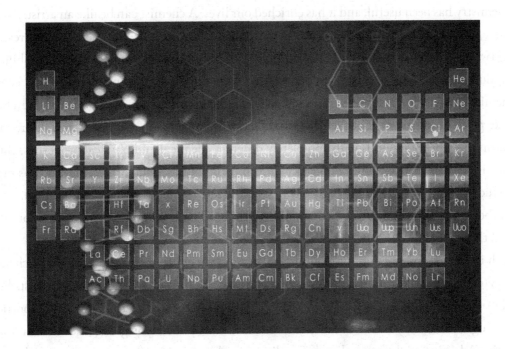

Periodic Tablet with DNA Strand

Chemistry is the study of matter and the physical and chemical changes it undergoes. A main approach in studying chemistry is relating the macroscopic world to the microscopic world. The details that go on in the microscopic world has always been an interest to many scientists. It is more than just curiosity. It is the mechanism and behavior of individual atoms, molecules, and ions that has been an interest by scientists for years. A lamp that has been turned on produces light. Rain gives off a rainbow. An iceberg floating on water and even a machine that is able to work has to do with chemistry. There are numerous examples that could be mentioned. The study of chemistry has even blended in with biology and physics as well as many other scientific fields and medical disciplines.

The materials that are made in this world has been carried out by chemical companies doing research. Silicone polymer chemistry has been useful for many applications such as drug-delivery. Pure chemicals and automatic chemical dispense systems have found applications to the semiconductor manufacturing process that is also used in the electronics industry. Alloys have been used for automobiles and aircrafts. In ceramic chemistry, manufacturers develop aqueous and non aqueous chemical synthesis to make ceramic powders. This is used for the fabrication of electronic components. In material science, a single atom laser shows classical laser and quantum mechanical properties. Emulsions, surfactants, emollients, moisturizers, waves, thickeners, active ingredients, sunscreens, color, and preservatives are ingredients used in cosmetic chemistry. Pharmacy is part of the health profession that connects the health sciences and the chemical sciences. Safe and effective use of pharmaceutical drugs is a goal for pharmacists. Radiological chemistry has to do with developing image enhancement agents that is used for therapy.

Chemistry has been useful, and it has enriched our lives. A chemist can be like an artist. An artist is creative in making many different kinds of paintings. Likewise, a chemist can also be creative in mixing two or more chemicals to produce unique products. Both professions find new combinations and produce original and beautiful products. Artists use paint to make pictures whereas chemists use chemicals to make brand new substances. Everything is made up of chemicals. The changes that take place in chemistry are called chemical reactions. When two or more chemicals are mixed, they form different compounds with completely different chemical and physical properties. But it is not just single elements combining together to form compounds. Huge molecules such as organic compounds can be synthesized too and put to use. In fact, cosmetics that are used on a daily basis contain both inorganic and organic compounds. We find that the development of compounds has changed the way we live.

Both inorganic and organic compounds are present everywhere on planet Earth. The chemistry of our planet such as the atmosphere and our oceans is constantly changing. Some scientists believe that global warming has already taken place. Sea levels are expected to rise and cover a portion of land throughout the world in the nearby future. How exactly society is going to deal with it is still uncertain. Adapting to different climate conditions by having cities float on oceans, underground cities, cities in mountains, and perhaps as floating space crafts or on the moon might be the only way to survive. In order to solve the scientific problems affecting our civilization, let us explore the details in the world of chemistry.

Chapter 2
Measurements

Beakers, Test Tubes, Flasks

The Scientific Method

Scientists are curious to know how things work. They use the scientific method which consists of making observations, formulating hypothesis, and carrying out experiments. There are two types of observations. Qualitative observations consist of what we see or feel. Quantitative observations consist of measurements. Measurements will be discussed later on, and they consist of a number and a unit. A hypothesis is formed to explain some type of experiments. Experiments test hypothesis. The scientist decides whether the hypothesis is valid or not. These three steps are repeated to get more results. Hypothesis that agrees with many observations are then made into a theory. Theories are also referred to as models. They consist of hypotheses that has already been tested and gives an explanation to new results. Theories can be changed. Experiments can be carried out and theories are refined. A natural law summarizes observed and measurable behavior.

Scientific Notation

Scientific notation is a specific notation that allows us to convert huge and small numbers into a digit between one and ten multiplied by its power of ten raised to a certain number. It is easier to handle numbers in scientific notation to avoid making a mistake in the decimal point or adding or subtracting a certain number of zeros. An example of scientific notation is

$$N \times 10^n$$

where N is a number between 1 and 10 and n is an integer. The decimal point is moved a certain number of places and this determines the power of ten. Moving the decimal point to the left gives a positive integer for the power of ten. Moving the decimal point to the right gives a negative integer for the power of ten.

Scientific Notation

(a) Express 734.267 in scientific notation

7.34×10^2

(b) Express 0.00000881 in scientific notation

8.81×10^{-6}

Arithmetic Operations

Adding and subtracting numbers involving scientific notation requires the exponent, n, to be the same. The N parts of the number can be added or subtracted.

Arithmetic Operations I

(a) $(6.3 \times 10^3) + (2.0 \times 10^3) = 8.3 \times 10^3$

(b) $(7.00 \times 10^4) + (3.0 \times 10^3) = 7.3 \times 10^4$

(c) $(3.00 \times 10^{-2}) + (5.0 \times 10^{-3}) = 3.5 \times 10^{-3}$

When we multiply numbers together in scientific notation, the N parts of the number are multiplied and the exponents, n, are added.

Arithmetic Operations II

(a) $(7.0 \times 10^{4}) \times (5.0 \times 10^{2}) = 3.5 \times 10^{7}$

(b) $(2.0 \times 10^{-5}) \times (8.0 \times 10^{-3}) = 1.6 \times 10^{-7}$

When we divide numbers together in scientific notation, the N parts of the number are divided and the exponents, n, are subtracted.

Arithmetic Operations III

(a) $(7.0 \times 10^{4})/(5.0 \times 10^{9}) = 1.4 \times 10^{-5}$

(b) $(8.0 \times 10^{6})/(3.0 \times 10^{-5}) = 2.7 \times 10^{11}$

International System of Units

Units always consist of two parts. These parts consist of a number and a unit. Numbers without units are meaningless. The unit is required because we want to know the size and magnitude of the number. Chemists normally use the units mass, length, time, temperature, electric current, and amount of substance. The English system of units is used in the United States of America. The metric system is used by most of the industrialized world. The United States has already started using the metric system. The International System, or SI system, bases itself on the metric system. Sometimes, prefixes are placed before the size of the unit. Units should be carried out in calculations. Finding the numerical answer to a solution without units is meaningless and sometimes can make the difference between life and death for a hospital patient. Units are measurements of a scale being used. Mass is normally measured in kilograms, and it is abbreviated as kg. Length is normally measured in meters, and it is abbreviated as m. Temperature is normally measured in degrees. Fahrenheit is abbreviated as °F. It is important to have it when carrying out measurements. There are several more units used in chemistry and in the other sciences. Changing one unit to another is always going to be important in all scientific fields and in life.

Prefixes Used with Units

Table 2-1: Prefixes and Their Order of Magnitude Used for Units

Prefix	Symbol	Meaning	Scale
Yotta	Y	10^{24}	Septillion
Zetta	Z	10^{21}	Sextillion
Exa	E	10^{18}	Quinillion
Peta	P	10^{15}	Quadrillion
Tera	T	10^{12}	Trillion
Giga	G	10^{9}	Billion
Mega	M	10^{6}	Million
Kilo	k	10^{3}	Thousand
Hecto	h	10^{2}	Hundred
Deca	da	10^{1}	Ten
Deci	d	10^{-1}	Tenth
Centi	c	10^{-2}	Hundredth
Milli	m	10^{-3}	Thousandth
Micro	μ	10^{-6}	Millionth
Nano	n	10^{-9}	Billionth
Pico	p	10^{-12}	Trillionth
Femto	f	10^{-15}	Quadrillionth
Atto	a	10^{-18}	Quintillionth
Zepto	z	10^{-21}	Sextillionth
Yocto	y	10^{-24}	Septillionth

Measurements

The length of a measurement is how long an object is from one end to another end. Rulers are normally the tools used to determine the length of an object. An example of a length measurement is 1 m = 100 cm. The volume of an object is the amount of three-dimensional space that a substance occupies. Glassware such as graduated cylinders, beakers, and erlenmeyer flasks are used to measure large volumes. An example of a volume measurement is the equivalence statement 1 L = 1000 mL.

The mass of an object is how much matter is in it. Electronic balances and older type of balances are used to measure the mass of an object. An example of a mass measurement is 1 kg = 1000 g.

Uncertainty in Measurements

Every measurement has some uncertainty in it. Measurements are never perfectly exact. If you consider the numbers 7.00 cm, 7.01 cm, 7.02 cm, and 7.01 cm, then you will notice the first two digits are the same. These are called the certain digits. The third digit is carefully estimated, and this is considered to be the uncertain digit. Whenever we make measurements, it is important to record all of the certain digits and the uncertain digits. Keep in mind it is possible their can be more than one uncertain digit. All measurements have some kind of uncertainty. Measuring devices such as pipettes, burettes, and electronic balances determine the uncertainty of a measurement. An example is measuring the volume of water using a graduated cylinder that has an uncertainty of 1 ml on its scale. If we measure the volume of a liquid to be 6 ml, for example, then the volume can be read between 5 ml to 7 ml. Another example is measuring the volume of water using a graduated cylinder with a much narrow scale and an uncertainty of 0.1 ml. If we measure the volume of a liquid to be 6.0 ml, then the real volume could fall in the range between 5.9 ml to 6.1 ml. Measuring the solution after the decimal place gives a more precise answer.

Significant Figures

Significant figures are specific numbers that have been recorded in a measurement, and they are the digits that contribute to precision. Significant figures can be abbreviated as s.f. They are digits that have meaning based on a measured or calculated quantity. Significant figures are the certain and uncertain numbers combined. Knowing the correct number of significant figures helps us determine how accurate and precise our measurement came out to be. The precision of significant figures is included except leading zeros, trailing zeros that are placeholders to indicate the scale of a number, and spurious digits which are digits carried out in calculations to a greater precision than the original data.

Guidelines

Nonzero integers are digits that are not zero are significant. For example, 894 has three significant figures because there are a total of three nonzero digits. The number 2.7341 has five significant figures because counting the digits from left to right, there are a total of five nonzero digits. Leading zeros are zeros to the left of the first nonzero digit, and these zeros are not significant. They are only used to show where the decimal point is located and to show the order of magnitude of the number. For example, the number 0.07 has only one significant figure. The nonzero digit, seven, to the very far right is the significant figure we have taken into consideration. The number 0.0000048 has two significant figures, and these digits include the four and eight.

Inserted or captive zeros are zeros between nonzero digits. For example, the number 707 has three significant figures. Counting the digits from left to right gives us a total of three digits that are significant and have meaning to them. The number 60,001 has five significant figures. The zeros in between the nonzero digits are counted. Without them, the number itself will have no meaning. Trailing zeros are numbers to the right of the decimal point if the digit to the left is a number greater than one. The number 3.0 has two significant figures. The uncertain digit zero to the right of the decimal point has meaning to it. The number 7.0000 has five significant figures. All of the zeros to the right are significant otherwise we would not take the time to write them out. The zero is precise to the ten thousands place.

Leading, inserted, and trailing zeros can be combined for a specific numerical measurement. Assuming the number is less than one, the zeros at the end of the number and zeros between nonzero digits are significant. Consider the number 0.00520 where it has both leading and trailing zeros. The number itself has a total of three significant figures. These numbers include the five, two, and the zero at the very end. If we take the number 0.00520 and write it in scientific notation, we get the number 5.20×10^{-3}. If you look at the N part of the number, you can see it has a total of three significant figures. The leading zeros were just placeholders to show the overall magnitude of the number.

Exact numbers are numbers that have an infinite number of significant figures. Numbers that do not have decimal points and that have trailing zeros may be significant or insignificant. Note that exact numbers are not measurements. Objects such as a ball, a pen, or a table are exact and are therefore considered to be an exact number.

Trailing zeros that do not contain a decimal point can be ambiguous. A bar can be placed over the last significant figure. Trailing zeros that follow are not significant. For example, the number 7100 with a bar over the first zero has three significant figures. Underlining the last significant figure for a number can also be done. For example, the number 3,000 with the first zero underlined has two significant figures. Placing a decimal point after a number could also indicate the proper number of significant figures. For example, the number 500. has three significant numbers. The number 700 has at least one significant figure. However, it is not clear how many exact number of significant figures there are. If we converted the number into scientific notation, we have three possibilities. The number could be 7×10^2, 7.0×10^2, or 7.00×10^3. If we take these numbers and convert it back to its expanded notation we get the final answer as 700. The best answer is to state the number 700 has at least one significant figure. For scientific notation, the numerical value of the representation that contains the significant figure, which does not include the base or the exponent, is called the significant or mantissa.

Rounding

Rounding numbers is very common in mathematics and in the sciences. Reducing our desired number to the proper number of digits is important for both accuracy and precision. If the digit to be removed is less than the number five, the digits before the number remain the same. Increase the number by one if the next digit is six or greater, or if there is a five followed by nonzero digits. If the last digit is a five followed by trailing zeros, then it should be changed to an even number. The rounded digit should increase if it is odd. Leave the digit alone if it is even.

The number 8.027 rounded to the hundredths place is 8.03 because the digit seven is six or more. Let us consider the number 8.023 rounded to the hundredths place. The answer is 8.02 because the next digit three is four or less. Moreover, the number 8.015 rounded to the hundredths place is 8.02 because the next digit is five and the digit in the hundredths place is one. But consider a bit of a more complicated rounding procedure a lot of people neglect. The number 8.045 is rounded to 8.04 to the hundredths place because the next digit is five and the hundredths digit is even. However, the number 8.04501 rounded to the hundredths place is 8.05 because the next digit is five and nonzero digits follow it.

We have to be careful when we round numbers in order to obtain both accurate and precise results. Whenever we carry out calculations, we want to carry extra digits all the way to the final answer. Once we obtain the final result, that is the time we round our measured number. When rounding a number, use the first number to the right of the last significant figure. Rounding off sequentially gives the wrong answer. If we consider the example 7.8145 rounded to three significant figures to the hundredths place, the answer is 7.81. The answer is not rounded to 7.815 then rounded to 7.82. Each digit used can make all the difference in the world. Representing a positive number k to a precision of p significant digits contains a value by the following formula

$$\text{Round}(10^{-n} \times k) \times 10^n \text{ where } n = \text{floor}(\log_{10}k) + 1 - p$$

Adding and Subtracting

The smallest number of significant figures to the right of the decimal point determines the correct number of significant figures. Decimal points are counted when adding and subtracting using significant figures. The number 80.442 + 1.1 is equal to 81.542 on your calculator. However, the number is correctly rounded to 81.5 because there is only one digit to the right of the decimal point. However, if we consider the number 1080 - 8.35, the answer becomes 1,072 since there is no decimal point to the right.

Multiplication and Division

The original number in the final calculation has the smallest number of significant figures. The measurement itself is limiting. For example, if we multiply 2.6 x 3.5029 the final answer on the calculator is 9.10754. However, the correct answer is rounded to only two significant figures. The answer is 9.1 since there are two significant figures in the original problem. Whenever we multiply and divide, we always count significant figures.

Exact Numbers

Exact numbers are considered to be objects that we can count or numbers obtained from definitions. For example, an object that weighs 2.214 grams and if there were eight objects, then the mass would be 2.214 grams x 8 which equals 17.71 grams. The number 8 is considered to be 8.00000…….by definition. This is why we don't round the final answer to one significant figure. Exact numbers do not have significant numbers since they are not measurements.

Logarithms

Finding significant figures using a base 10 logarithm for a normalized number is done differently. The result is rounded to the number of significant figures for the normalized number. For example, consider the following computation

$$\log_{10}(2.0 \times 10^5)$$

$$\log_{10}(2.0) + \log_{10}(10^5)$$
$$0.301029996 + 5$$

$$5.30$$

The final answer is rounded to 5.30 since the last two digits .30 has the same number of significant figures as 2.0 in the original number. The value 0.301029996 is the mantissa and the value 5 is the characteristic. When taking the antilogarithm of a number, the final number should have the same significant figures as the mantissa of the logarithm. When carrying out a calculation, never round for intermediate results. Keep at least one more digit than the precision of the final result. Always round at the end of a calculation in order to avoid cumulative rounding errors. This will help you keep the answer both accurate and precise as possible. In fact, this is what we strive for in chemical calculations.

Dimensional Analysis

Converting one unit of measurement to another is done by conversion factors. We call this Dimensional Analysis or The Factor Label Method. For example, if we consider the equivalent statement 1 dollar = 10 dimes, they represent the same amount but the units dollar and dimes are different. If we were to rewrite the equivalence statement in terms of a ratio, then the ratio could be either 1 dollar/10 dimes or 10 dimes/1 dollar. These ratios are called conversion factors. Both the numerator and denominator tell us the same amount. We use conversion factors to carry out conversions with different units. When carrying out conversion factors, we use one or more conversion factors based on equivalent statements. We multiply one or more conversion factors to give us the desired unit at the very end. At the end of the calculation, check to see if you have the correct number of significant figures and make sure the answer makes sense. If you set up the equation properly, then all the units should cancel except the unit at the end.

Temperature

Temperature is used a lot in the sciences. It is important to know if an object is hot, warm, cold, or very cold. Knowing the temperature of a patient can mean the different between life and death. The equation for temperature to convert from degrees Celsius to Kelvins is

$$T_K = (T_{oC} + 273) \times (1 \text{ K}/1°\text{C})$$

Other equations used to calculate temperature are as follows:

Fahrenheit [°F] = [°C] x (9/5) + 32

Kelvin [K] = [°C] + 273.15

Rankine [°R] = ([°C] + 273.15) x (9/5)

Romer [°RO] = [°C] x (21/40) + 7.5

Newton [°N] = [°C] x 33/100

Delisle [°De] = (100 - [°C]) x (3/2)

Reaumer [°Re] = [°C] x (4/5)

Density

Density is the amount of matter that is in a volume of a substance. Some objects are really dense and can sink at the bottom of a volume of liquid where other objects are less and can either float in solution or float on the surface of a solution. The equation for density is

$$\text{Density} = \text{mass}/\text{Volume}$$

Specific Gravity

Specific Gravity is defined as comparing the density of a liquid to the density of water. If the density of the liquid is larger than the density of water, then that means water is less dense than the liquid. We can compare other states of matter such as a gas. If the density of a gas is larger than the density of air, then that means air is less dense than the gas. The equation for specific gravity is

$$\text{S.G.} = \text{density of a liquid}/\text{density of water at } 4°\text{C}$$

The equation can also be rewritten as

$$\text{S.G.} = \text{density of a gas/density of air}$$

Specific gravity or relative density for a gemstone is determined by comparing the weight of the gemstone with the weight for an equivalent amount of water. Measuring the specific gravity of a gemstone is based on Archimede's Principle. The apparent loss of weight for an object when submerged in water equals the weight of the water displaced. The weight of the water displaced equals the weight of the gemstone if it was made of water. The gemstone can be weighed in air and then in water. The specific gravity for the gemstone is the weight of the gemstone in air divided by the weight in air minus water

$$\text{S.G.} = \text{Air/(Air - Water)}$$

In other words, we are taking a ratio of two quantities. Note that specific gravity is unitless.

Conclusion

The scientific method is used to make or develop theories and then maybe into laws. Quantitative measurements consist of a number and a unit. There is always some uncertainty associated with each measurement. Significant figures or significant digits consist of certain and uncertain digits. The English and metric system is used to convert units to each other. Converting units to another is called Dimensional Analysis of The Factor Label Method. There are many equations used to convert one unit of temperature into another. The most common conversions involve changing degrees Celsius to Kelvins, Kelvins to degrees Celsius, degrees Celsius to degrees Fahrenheit, and degrees Fahrenheit to degrees Celsius. Density is defined as mass per unit of volume. Specific gravity is the ratio of the density of a liquid to the density of water. It can also be expressed as the ratio of the density of a gas to the density of air. There are no units for specific gravity. The topic of measurements and its calculation is important, and it will be seen throughout the remainder of the material covered.

Chapter 3
Atomic Structure

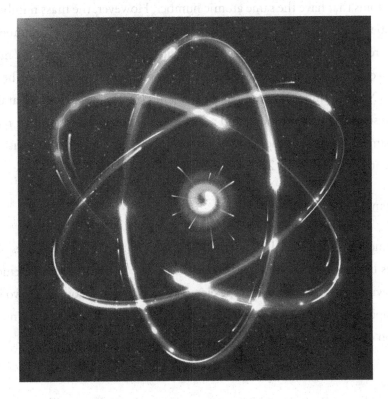

Shining Atom Vector Neon Scheme

Dalton's Atomic Theory consist of simple statements. He considered elements as small particles which he named them atoms. Atoms of a same element are the same and identical, but they are different from atoms from different elements. It is possible that atoms of one element combine with atoms of another element to make compounds. Atoms are also not seen during chemical processes.

We cannot create or destroy them when chemical reactions have taken place. The only thing that is changed in atoms is that they are regrouped.

Let us look deeper into the atom. An atom can be thought of as a unit that enters in chemical reactions. They have a unique internal structure made up of subatomic particles. The subatomic particles mainly used in chemistry are the proton, the electron, and the neutron. The protons and neutrons are held together in a central core called the nucleus. If there is more than one nucleus, we call them nuclei. Protons which have the symbol p are particles that are positively charged with an extremely small mass. The charge of a proton can be written either as 1+ or +1.

Newtons which have the symbol n are particles that are neutral with a slightly larger mass than the proton. The charge of a neutron is zero. Electrons are denoted as the symbol e^- and they consist of negatively charged particles. The mass of an electron is a lot smaller than the mass of a proton with a charge of 1 - or - 1. The atom itself has a small nucleus that consist of protons and neutrons. Electrons surround the nucleus. Keep in mind each atom have different numbers of protons, neutrons, and electrons.

Isotopes are atoms that have the same atomic number. However, the mass numbers are different. The atomic number is the number of protons inside the nucleus. The mass number consist of the total number of protons and neutrons inside the nucleus. If we subtract the mass number from the atomic number, we can determine the number of neutrons inside the nucleus. The symbol for the isotope is given as X. The symbol X represents an element. The symbol A is given at the upper left of the X which represents the mass number, and a Z is given at the lower left which represents atomic number. Mass number is equal to the number of protons plus the number of neutrons. Solving for the number of neutrons gives us A - Z.

Elements and Compounds

Before we name specific chemicals, it is important to know and learn the names of specific solids, liquids, and gases that contain an entity called elements. The periodic table of elements contain all the elements known to exist and not exist. The elements are known to consist as two letters. The first letter is always capitalized and the second letter is never capitalized. The names and the symbols of the most common elements are as follows

Table 3-1: Background Information About the Common Elements

Name	Symbol	Discoverer	Date of Discovery
Silver	Ag	Unknown	Ancient
Aluminum	Al	F. Woehler	1827
Argon	Ar	Albertus Magnus	1250

Table 3-1: Background Information About the Common Elements

Name	Symbol	Discoverer	Date of Discovery
Gold	Au	Unknown	Ancient
Boron	B	Sir Humphry Davy J.L. Gay-Lussac L.J. Thenard	1808
Barium	Ba	Sir Humphry Davy	1808
Bismuth	Bi	Claude Geoffroy	1753
Carbon	C	Unknown	Ancient
Calcium	Ca	Sir Humphry Davy	1808
Cadmium	Cd	Fr. Stromeyer	1817
Chlorine	Cl	K.W. Scheele	1774
Cobalt	Co	G. Brandt	1735
Chromium	Cr	L.N. Vauquelin	1797
Copper	Cu	Unknown	Ancient
Fluorine	F	H. Moissan	1886
Iron	Fe	Unknown	Ancient
Hydrogen	H	Sir Humphry Davy	1766
Helium	He	P. Janssen Sir William Ramsey	1868
Mercury	Hg	Unknown	Ancient
Iodine	I	B. Courtois	1811
Potassium	K	Sir Humphry Davy	1807
Lithium	Li	A. Arfvedson	1817
Magnesium	Mg	Sir Humphry Davy	1808
Manganese	Mn	J.G. Gahn	1774
Nitrogen	N	Daniel Rutherford	1772
Sodium	Na	Sir Humphry Davy	1807
Neon	Ne	Sir William Ramsey M.W. Travers	1898
Nickel	Ni	A.F. Cronstedt	1751

Table 3-1: Background Information About the Common Elements

Name	Symbol	Discoverer	Date of Discovery
Oxygen	O	Joseph Priestly C.W. Scheele	1774
Phosphorus	P	H. Brandt	1669
Lead	Pb	Unknown	Ancient
Platinum	Pt	Charles Wood A. de Ulloa	1735
Radium	Ra	Pierre Curie Marie Curie	1898
Sulfur	S	Unknown	Ancient
Antimony	Sb	Unknown	Ancient
Silicon	Si	J.J. Berzelius	1824
Tin	Sn	Unknown	Ancient
Strontium	Sr	Sir Humphry Davy	1807
Titanium	Ti	W. Gregor	1791
Tungsten	W	J.J. de Elhuyar F. de Elhuyar	1783
Uranium	U	M.H. Klaproth E.M. Peligot	1789 1841
Zinc	Zn	A.S. Marggraf	1746

Given the name of the element, we can write a chemical formula. A chemical formula shows the chemical composition of a compound. The symbols of the elements are shown above. The element symbol itself represents an atom. A subscript is given at the right of the element symbol to represent the number of each type of atom. A subscript of one is not written if there is only one atom. The Periodic Table is a chart that shows the arrangement of the elements. An element is considered to be a substance that cannot be separated into simpler substances. Each of the elements are arranged in terms of increasing atomic number.

The elements are mainly arranged due to similarities in chemical properties. Each of the elements arranged in a vertical column on the periodic table are called groups. Each of the elements arranged in a horizontal row on the periodic table is called a period. The first column towards the left hand side is called the Alkali Metals. The second column is called the Alkaline Earth Metals. Then there is a middle block on the periodic table. The middle block is considered to be the transition metals.

The elements towards the right hand portion of the periodic table are called noble gases. The second to the right hand column of the periodic table are called halogens.

The periodic table is also categorized into three categories. These categories are metals, nonmetals, and metalloids. The majority of the elements on the left hand side, middle, and a small portion on the bottom right hand side are considered to be metals. Metals are considered to be good conductors of heat and electricity. Metals provide good conduction of both heat and electricity. They can be hammered into sheets, and they can be pulled into wires. The neat feature about metals is that they have a shiny appearance.

Nonmetals are considered to be poor conductors of heat and electricity. Their chemistry is a lot different from the metals themselves. Metalloids are considered to be semimetals, and they fall between the characteristics between metals and nonmetals. If we look closely at the periodic table, we see the elements gold, silver, and platinum. These are considered to be not reactive. They are referred to as noble metals. The nonmetallic elements at the very end column of the periodic table are called noble gases. Seven nonmetallic elements are considered to be diatomic. The diatomic molecules are nitrogen, oxygen, hydrogen, fluorine, chlorine, bromine, and iodine. They are considered to be gaseous at normal temperatures. However, bromine and mercury are considered to be liquids at normal temperatures.

The natural states of the elements also consist of allotropes. Allotropes consist of the same element with different forms, but they each leave different properties. Examples are diamond, graphite, buckminsterfullerene, and carbon nanotubes. We have talked about elements, but these elements can be charged. Elements that lose or gain an electron are considered to be ions. If the ion is positively charged, it is considered to be a cation. This means one of more electrons was removed from a neutral atom. If the ion is negatively charged, it is considered to be an anion. This means one or more electrons were added to the ion. Ions in the first column of the periodic table have a charge of +1. Ions in the second column of the periodic table have a charge of +2. Normally, ions in the third column of the periodic table have a charge of +3. Elements towards the top and third to the last column of the periodic table have a -2 charge. The second to the last column have a -1 charge. The elements in the middle of the periodic table in which we referred to as transition metals have a wide variety of charges.

We can actually combine cations and anions together. When we combine cations and anions together we form a compound. A compound is a substance that consist of two or more elements combined together. The Law of Definite Proportions or The Law of Constant Composition states that when you take different samples of a compound, the elements will have the same proportions by mass. For example, the water molecule will always have two hydrogen atoms and one oxygen atom whether it is from the planet Earth or some nearby planet. The Law of Multiple Proportions states that two elements combined to form more than one compound, the masses of one element combine a known mass of another element have ratios of small numbers. For example, two stable compounds such as sulfur dioxide and sulfur trioxide have small ratio numbers.

Compounds containing ions have high melting points. If a compound dissolves in water, it breaks apart into the cations and anions. When the ions are dissolved in water, the solution can conduct electricity. Consider table salt that is common in chemistry laboratories. An example of table salt is sodium chloride. Sodium chloride in its solid state does not conduct electricity. However, when it is dissolved in water, it conducts electricity because the ions are charged. If we take a sample of pure water, we find no electricity is conducted because there are no ions present in solution. This shows that elements and compounds have different characteristics. Let us take a look at the cations and anions that make up different compounds. The cation is normally the metal and the anion is normally the nonmetal. The compound itself must have an overall charge of zero. Chemical compounds are neutral.

Matter can be defined as everything that occupies space and has mass. There are three states of matter. There is the solid state, the liquid state, and the gaseous state. Solids are rigid and have a fixed shape and volume. Liquids have a definite volume and at the same time it assumes the shape of the container itself. Gases have fixed volumes and shapes, and they assume both the shape and the volume of the container. Each of the states of matter can be described by its physical and chemical properties. A physical property is a property that is either measured or observed, but the identity of the substance itself does not change. Examples of physical properties include matter, volume, color, odor, melting point, boiling point, or density. A chemical property, however, is a property where a chemical change takes place. Brand new substances are made. Examples of chemical properties include iron rusting, plants changing color, gasoline being combusted, or metabolizing the food we eat.

Let us take a look at physical and chemical changes. Physical changes involve changes in the physical properties, but the component itself remain the same. Examples include reversible processes such as a solid, liquid, and a gas. Chemical changes involve changes in components for different kinds of substances. Substances can change into different kinds of substances. Chemical changes are often referred to as reactions. Matter has measurable properties. Values that measure the amount of matter are called extensive properties. Extensive property values can be added together. Examples include adding values of mass, length, or volume. Intensive properties consist of measured values, and they do not depend on the amount of matter that is present. Examples of intensive properties include color, melting point, solubility, acidic or alkaline solutions, density, or chemical composition.

A mixture consists of a combination of two or more substances, but the identity of the substances are the same. Mixtures can be separated into their original substances whether it is an element or a compound. It is possible for them to be converted into more than one kind of substance. Homogeneous mixtures have the same composition throughout the solution after stirring. Heterogeneous mixtures consist of visible components that can be separated from each other. A pure substance has a consistent composition, but unique properties are different from other substances.

Mixtures can be separated from each other. Distillation is one of the processes used to separate two liquids with different boiling points. Vaporizing a liquid in a boiling flask and then condensing it to another flask is the basic process of distillation. Filtration is another technique that is used to separate a solid from its liquid mixture. The heterogeneous mixture consist of both a solid and a liquid that is poured onto a porous barrier where the liquid passes through. The solid is left onto the porous paper and the liquid that is collected through the funnel is called the filtrate. Another purpose of filtration is to remove any impurities from the liquid. Filtration is common in chemistry laboratories, the healthcare profession, and even water technology. It is a useful technique carried out to get the job done.

Nomenclature

There are so many chemical compounds. To name inorganic compounds require that we learn a set of rules and conditions. We can classify inorganic chemical compounds into three types: I, T, and N. Binary compounds consist of two elements. These can be compounds that consist of only metals and nonmetals, whereas there are other compounds that contain only nonmetals. Let us take a look at Type I compounds. These compounds are considered to be binary ionic. Type I compounds consist of mainly the Alkali, Alkaline Earth Metals, and sometimes Aluminum as a cation. The anion is normally the nonmetal. Remember the Alkali Metals have a +1 overall charge and the Alkaline Earth Metals have a +2 overall charge. Their charges are fixed. For example, the metal sodium combines with the nonmetal chlorine. The metal magnesium combines with the nonmetal oxygen. When we name Type I binary ionic compounds, we always name the cation and then the anion. The cation is the same as the name of the element. The anion ending has the suffix "ide". Metals and nonmetals combined together are considered to be ionic and the overall charge must be zero. Roman numerals are not needed for Type I binary ionic compounds because their charges are fixed.

Type T binary ionic compounds consist of mainly transition metal elements. They serve as the cations for nonmetallic elements. Transition metals have various charges. Roman numerals are needed to show the oxidation state. Type N Compounds are not considered to be ionic. They involve the use of nonmetals. The first element is the cation and the element name is used. The name of the second element is the anion itself. We use prefixes to express the number of atoms present. Normally, the prefix "mono" is never used when naming the first element. The following are prefixes used to show the numbers in chemical names.

Table 3-2: Prefixes Used for Compounds

Prefix	Number
mono	1
di	2
tri	3
tetra	4
penta	5

Table 3-2: Prefixes Used for Compounds

hexa	6
hepta	7
octa	8
nona	9
deca	10

There are also ions that consist of one or more atoms. Polyatomic ions have more than one kind of atom. Oxyanions have elements combined with different number of oxygen atoms. Oxyacids are acids that have H, O, and other elements. If we take a representative acid that ends as "ic" and if we add an oxygen atom we get the prefix and suffix "per…..ic" acid. But if we remove an oxygen atom from "ic" acid, we get "ous" acid and by removing another oxygen atom we get "hypo…..ous" acid. If we go back and consider the "ic" acid and remove H^+ ions we get "per….ate". If we remove an oxygen atom from "ate" we get "ite". Removing another oxygen atom from "ite" gives us the prefix and suffix "hypo……ite". Examples of these kinds of acids and anions are as follows

Table 3-3: Common Oxyacids and Oxyanions

Oxyacid	Chemical Name	Oxyanions	Chemical Name
$HClO_4$	Perchloric Acid	ClO_4^-	Perchlorate
$HClO_3$	Chloric Acid	ClO_3^-	Chlorate
$HClO_2$	Chlorous Acid	ClO_2^-	Chlorite
$HOCl$	Hypochlorous Acid	OCl^-	Hypochlorite

Table 3-4: Most Common Polyatomic Ions

Chemical Name	Ion	Chemical Name	I
Ammonium	NH_4^+	Fluoride	F^-
Acetate	$C_2H_3O_2^-$	Hydride	H^-
Arsenate	AsO_4^{3-}	Hydroxide	OH^-
Bicarbonate	HCO_3^-	Hypochlorite	ClO^-
Bisulfate	HSO_4^-	Iodate	IO_3^-
Bisulfite	HSO_3^-	Iodide	I^-

Table 3-4: Most Common Polyatomic Ions

Chemical Name	Ion	Chemical Name	I
Borate	BO_3^{3-}	Nitrate	NO_3^-
Bromate	BrO_3^-	Nitrite	NO_2^-
Bromide	Br^-	Oxalate	$C_2O_4^{2-}$
Carbonate	CO_3^{2-}	Oxide	O^{2-}
Chlorate	ClO_3^-	Perchlorate	ClO_4^-
Chloride	Cl^-	Permangante	MnO_4^-
Chlorite	ClO_2^-	Peroxide	O_2^{2-}
Chromate	CrO_4^{2-}	Phosphate	PO_4^{3-}
Cyanide	CN^-	Phosphide	P^{3-}
Dichromate	$Cr_2O_7^{2-}$	Phosphite	PO_3^{3-}
Silicate	SiO_3^{2-}	Superoxide	O_2^-
Sulfate	SO_4^{2-}	Sulfide	S^{2-}
Sulfite	SO_3^{2-}	Thiocyante	SCN^-

We can also name a certain type of inorganic compound called acids. Acids are molecules that can give hydronium ions (H^+) when they are dissolved in water. If oxygen is not present in the compound, the prefix "hydro" and the suffix "-ic" is added to the name of the compound. When oxygen is present, the suffix could be -ic or -ous. If the anion ends in -ate, the suffix -ic is used. If the anion ends in -ite, the suffix -ous is used.

Conclusion

Atoms have a nucleus that contains protons and neutrons that are surrounded by electrons. The mass of an electron is small, and the charge of an electron is even smaller. Protons have a positive charge, and neutrons have no charge. Isotopes have the same number of protons, but the number of neutrons is different. In other words, isotopes have different mass numbers, but their atomic numbers are the same. Atoms also form molecules by forming covalent bonds. They do this by forming covalent bonds. A molecule can be represented by a chemical formula, a structural formula, by space-filling models, by ball-and-stick models, or it can even be shown on a computer screen using a CD-ROM. Cations are atoms that lose one or more electrons. Ionic compounds contain charged ions. Elements are arranged based on atomic number that is increasing. The groups or families on the periodic table have chemical properties that are almost similar to each other. Normally, most metals form cations with nonmetals to make anions. Metals and nonmetals combine with each other to form ionic compounds. Binary ionic compounds contain a metal that is followed by the root name of the nonmetal. Metals containing cations with different charges are indicated using a Roman numeral. Binary covalent compounds contain nonmetals. Prefixes are used for binary covalent compounds. They tell us how many atoms are present.

Chapter 4
Stoichiometry

Laboratory Equipment

Assume you have a sample of peanuts that has a mass of fifty grams. A friend of yours asks for five hundred peanuts. It would take a long time to count five hundred peanuts. Instead, it would be better to multiply five hundred peanuts by the ratio of one sample per fifty grams to obtain ten grams. It is a lot easier to weigh out ten grams on an electronic balance rather than to count five hundred individual peanuts. Keep in mind that each of these peanuts is not going to be the same size and their masses will be slightly different. But that is not a problem because if we take an average mass of a group of peanuts, we would get according to the information above ten grams. The peanuts can be treated like a sample as long as we assume they are all identical.

The point is that objects do not need to be the same masses to be counted by the method of weighing them. You just need to know the average mass of the objects. Objects can behave as being the same for the method of counting. Suppose your friend wanted the same number of peanuts and walnuts. You already know a sample of peanuts has a mass of fifty grams. You also have a mass of walnuts to be one hundred grams. You weigh out a mass of peanuts and put them on the electronic balance that reads one thousand grams. Let us look at how can determine to have the same number of walnuts and peanuts if you have one thousand grams of peanuts.

Notice that there is twice the amount of walnuts as there are peanuts. In other words, you would have to weigh out 2 X 1,000 grams to obtain 2,000 grams of walnuts. The point is that two samples that have different components can have the same number of components if the ratio of sample masses is the same ratio of the individual components. In other words, samples that have the same mass ratio will have the same number of components. Consider the reaction

$$2Na(s) + Cl_2 \rightarrow 2NaCl(s)$$

and that this reaction will go to completion in the forward direction. We can weigh out the mass of sodium since it is a solid. But we cannot weigh out the mass of chlorine since it is a gas. We would not know the number of chlorine molecules directly by looking at the equation either. We would have to find an indirect method to determine the number of chlorine gas molecules. We already know the mass of molecules, atoms, and ions is very small. A smaller unit of mass called the atomic mass unit or amu is sometimes used in calculations. It has the equivalence statement 1 amu = 1.66×10^{-24} grams. If we are going to determine the number of molecules, atoms, and ions, we would need to look at the atomic mass unit of that specific entity.

If we look at the periodic table of elements, we find there are 14.01 amu for every one nitrogen atom. If we wanted to determine the number of nitrogen atoms in a sample of 28.02 amu, we would take the ratio 28.02 amu over 14.01 amu to obtain two atoms. You notice that the numerical values of these samples are in decimals. They are not exact numbers. Each element on The Periodic Table consist of a mixture of many isotopes. The atomic numbers are the same, but the mass numbers of each isotope is slightly different. Knowing the mass of a sample and the average mass, we can determine the number of atoms, molecules, or ions. Average atomic masses are shown below on The Periodic Table. Sometimes, they are called the atomic weights. The term mass number is used when we discuss protons and neutrons. The term atomic weight is used when we discuss atomic masses.

Suppose we have 16.00 grams of oxygen atoms and 14.01 grams of nitrogen atoms. Let us discuss how we would determine the same number of atoms. If we take the ratio of the masses in grams which is the same numerical value of the masses in amu, we would find the masses of each atom will always have the same number of atoms. Weighing the samples out of each element that is equal to the average atomic mass, we find these samples have the same number of atoms. We use the term mole to represent the atomic weight of each element. The abbreviation of a mole is mol. It is equal to 6.022×10^{23} molecules, atoms, or ions. Recall that this numerical value is Avogadro's Number. It is not necessary to memorize atomic weights. Normally, they are given on quizzes and examinations.

One of the goals in chemistry is to study chemical changes. Stoichiometry, as we shall study later, is based on chemical reactions and how much material is being used and how much is being made. The atomic weight for carbon, for example, is 12.01 grams. It is calculated based on average atomic masses of carbon and its corresponding percentages. By weighing a sample of carbon, we are able to count the number of atoms. The unit for mole is used for counting atoms. A mole equals the number of carbon atoms in 12 grams of pure carbon. This number is called 6.022×10^{23} units. A mole of a quantity contains this many units of a substance. We can also define a mole that contains a sample of an element with a mass equal to the atomic mass of an element that contains 1 mole of atoms.

The molar mass of a substance is obtained by taking the sum of the masses of each of the components. It tells us the mass of a mole of a certain element, molecule, compound, or ion. If we wanted to determine the molar mass of table salt which is NaCl, then we would look up on the Periodic Table of Elements the masses of sodium and chlorine. We would find the mass of sodium to be about 23.0 grams/mole, and the mass of chlorine to be 35.45 g/mol. If we take the sum of these two numerical values, we would obtain 58.45 g/mol. This represents the molar mass of sodium chloride. We can think of a compound as having many atoms. The molar mass of a molecule or compound represents the sum of the masses of 1 mole of each of its elements. Molar mass is the mass in grams of a mole of a compound. The term molecular weight has also been used to describe molar mass. To calculate molar mass of water, the mass of 2 mol H is 2.016 grams, and the mass of oxygen is 16.00 grams added together to give 18.016 grams.

If we take a look at the previous example, we found the molar mass of the compound for carbon dioxide. The formula for carbon dioxide is CO_2. This formula tells us that there is 1 mole of carbons and 2 moles of oxygens. It also tells us that there is 1 atom of carbon and 2 atoms of oxygens. There are over millions of formulas of chemical compounds and molecules. There is no need to memorize all of them instead it is important to understand what a formula tells us. We can use formulas to determine the percent composition of each of the elements. If we know the mass of each element divided by the mass of 1 mole of the compound and then multiply it by 100%, we can obtain the mass percent of that element. This is the same idea when you get a score on a test and divide by its total and then multiply by 100%.

Compounds can be described by the number of atoms and based on its percentage of elements. The goal of percent composition is to calculate the mass percents of the elements. This is done by taking the subscripts and multiplying it by the atomic weights of each element, and then summing the values for each element to obtain the total mass of one compound. Then the mass percent, sometimes called the weight percent, is obtained by comparing the ratio of the mass of an element to the total mass of a compound. The result again is then multiplied by 100%.

Mass Percent = Molar Mass of the Element/Molar Mass of the Compound

The chemical formulas shows the number of atoms for a given compound. The mass of the elements is converted to atoms. Calculating the empirical formula of a compound consist of a series of steps. The percentage of each element is converted to its mass that is typically in grams. Each element is then converted to moles using atomic masses. The values are divided by the smallest value computed for the element. The results should represent a whole number after rounding. The numerical values are the subscripts for the empirical formula. If the resulting numerical values are not whole numbers, then each value is multiplied by an integer. The empirical formula is considered to be the simplest formula that shows the whole number ratios for each of the atoms in a compound or molecule. The molecular formula, however, is the real or actual formula that shows the composition of the compound or molecule. To determine the empirical formula of a compound, the mass of each of the elements is obtained by converting all percentages to grams. The molar mass for each of the elements is then used to convert the masses to moles. The moles are divided by the smallest number of moles for each of them. The numbers obtained should be whole numbers. If these numbers are not whole numbers, then multiply each of them by a number to obtain whole numbers. The whole numbers are the subscripts for the empirical formula.

To determine the molecular formula, we need to know the empirical formula and the molar mass of the compound or molecule. In other words, the following holds true. There are two methods for calculating the empirical formula. We want to obtain the empirical formula. The mass should be converted according to the empirical formula. The ratio of molar mass to empirical formula mass should be computed. The integer tells us the number of empirical formula units for a molecule. The empirical formula subscripts are then multiplied by this integer. A molecular formula is then obtained. Another method involves finding out the mass of an element in a mole of a compound. This is carried out using the mass percentages and the molar mass. The next step is finding the moles of each element on one mole of a compound. The integers tell us the subscripts in the molecular formula.

Molecular Formula = (Empirical Formula)$_n$

where n is considered to be a whole number

n = Molar Mass/Empirical Formula Mass

Stoichiometry involves determining the amounts of reactants and products based on a balanced chemical reaction. Given the amount of mass in the reactants or products, we can determine the number of moles for that specific molecule or compound. Using ratios based on the stoichiometric coefficients in the balanced equation, we can determine mole-mole ratios for the desired reactant or product. Once the final number of moles is determined, then using the molar mass of the desired molecule or compound, we can calculate the final mass of the reactant or product. Consider the following equation

$$CH_3OH(l) + O_2(g) \longrightarrow CO_2(g) + 2H_2O(g)$$

This equation can be determined as 1 mole of liquid methanol reacts with 1 mole of gaseous oxygen to give 1 mole of gaseous carbon dioxide and 2 moles of gaseous water. Chemical equations such as this can be used to calculate masses of reactants or products. The first step will be to balance the chemical equation. If the mass of the reactants is given, then the number of moles should be calculated using the molar mass of the element, compound, or molecule. The balanced equation is used to set up mole ratios, and these derived mole ratios is used to calculate the reactant or product.

If equal quantities of reactants are mixed together, they will run out at the same time. This means the reactants are mixed in stoichiometric quantities. Sometimes, the reactants are mixed in unequal proportions. The reactant mixed in with the smaller proportion will be consumed first. This is called the "limiting reactant". The other reactant is considered to be present in excess, and we call it the excess reactant.

The amount of products calculated from the limiting reactant is normally considered to be the theoretical yield. The actual yield is the amount of product obtained in the laboratory. It is very rare for the actual yield to be exactly the same as the theoretical yield since the majority of reactions do not go to completion. The formula for percent yield is given below.

$$(Actual\ Yield/Theoretical\ Yield) \times 100\% = Percent\ Yield$$

Limiting Reagent

The coefficients according to a chemical equation tell us the number of molecules. We are interested finding out the mass of either the reactants or products. Make sure the equation for the reaction is balanced before carrying out stoichiometric calculations. It is important to convert between masses and moles of substances. Mole ratios are being used to carry out stoichiometric calculations. Finally, we want to convert moles back to mass in grams if this is what the problems asks us. Sometimes, exact amounts of reactants are used and mixed together based on stoichiometric quantities. Sometimes, when two chemicals are mixed together, one of them is present in a smaller amount, and the other one is present in a greater amount. Suppose, for example, Molecule A contains 30.0 grams and Molecule B contains 60.0 grams. Then Molecule A will get consumed first, and Molecule B will remain in excess perhaps at least 30.0 grams. We can say that Molecule A is the limiting reagent. The formation of the product depends on the limiting reactant. It is important to know what the limiting reagent is so we can calculate the amount of product. Assume there are 4 H_2 and 1 O_2 molecules placed into a container. According to the chemical equation,

$$2 \, H_2 + O_2 \longrightarrow 2 \, H_2O$$

we see that it takes 2 moles H_2 and 1 mole O_2 to react to form 2 moles H_2O. If we set up a ratio 1 O_2/2 H_2, we say that it takes 0.5 O_2/1 H_2. In our container, we have

$$1 \, O_2/4 \, H_2 = 0.25 \, O_2/1 \, H_2.$$

This is the actual ratio. However, the required ratio is 0.5 O_2/1 H_2. What we find here is that O_2 is the limiting reactant.

Let us look at the equation in a different way. Suppose we have the equation

$$2 \, H_2 + O_2 \longrightarrow 2 \, H_2O$$

and we decide to multiply the equation by 2. We would get

$$4 \, H_2 + 2 \, O_2 \longrightarrow 4 \, H_2O.$$

This equation states that we have 4 moles H_2 that require us to react with 2 moles to form 4 mol H_2O. However, we only have 1 mole O_2 in our container. This is another way to look at O_2 being the limiting reactant.

Suppose there are 5.0 grams O_2 and 1.22 grams H_2, how much H_2O is formed?

Converting (5.0 grams O_2)(1 mol O_2/32.0 grams O_2) = 0.16 moles O_2

Converting (1.22 grams H_2)(1 mol H_2/2.016 grams H_2) = 0.605 mol H_2

Converting (0.16 mol O_2)(2 mol H_2/1 mol O_2) = 0.32 mol H_2

then this means this is the amount that is required to react with 0.16 moles O_2. However, we have 0.605 mol H_2, and this tells is that H_2 is present in excess and O_2 is the limiting reagent.

Here is another method to look at the problem. Again the equation is

$$2H_2 + O_2 \longrightarrow 2H_2O$$

Converting (5.0 g O_2)(1 mol O_2/32.00 g O_2)(2 mol H_2O/1 mol O_2)(18.016 g H_2O/1 mol H_2O) = 5.63 g H_2O

Converting (1.22 g H_2)(1 mol H_2/2.016 g H_2)(2 mol H_2O/2 mol H_2)(18.016 g H_2O/1 mol H_2O) = 10.9 g H_2O

Since the smallest number is 5.63 g H_2O, we can conclude that O_2 is the limiting reagent.

Another method is as follows. Comparing mole ratios of elements from the balanced equation to the mole ratio of reactants that are present gives us

$$1 \text{ mol } O_2/2 \text{ mol } H_2 = 0.5 \text{ mol } O_2/1 \text{ mol } H_2$$

This is the required ratio. Comparing the ratio

$$(0.16 \text{ mol } O_2/0.605 \text{ mol } H_2 = 0.26 \text{ mol } O_2/1 \text{ mol } H_2)$$

This gives us the actual ratio. Since 0.26 moles is less than 0.5 moles O_2, we find that O_2 is the limiting reactant or limiting reagent. The amount of product formed based on the calculation is the theoretical yield. We can think of this as the maximum amount of product that is formed. Normally, the amount of product formed never reaches the theoretical yield according to experimental conditions. The yield given in the laboratory experiment is the actual yield. We can calculate the percent yield as taking the ratio of actual yield to theoretical yield and multiplying it by 100%. Carrying out this calculation gives us the percent yield. When solving a stoichiometry problem, we have to keep some things in mind. Make sure the chemical equation is balanced. The mass should also be converted to moles. The limiting reactant should be determined. Mole ratios should be set up and the number of moles of product should be formed. Moles should finally be converted to grams using molar mass.

Conclusion

Stoichiometry has to do with the amount of materials involving reactions. Average atomic mass is obtained taking the average of different masses and percentages of its isotopes. A mole is considered to be 6.022×10^{23} units. This number is Avogadro's number. We can think of the elements as having atomic masses expressed in grams/moles which are the units for molar mass. It is obtained by taking the sum of each of the elements to obtain a value. We can also describe the composition of each of these elements such as percent composition and empirical formulas. Mass percent is the mass of an element divided by the total mass of the substance multiplied by 100%. The empirical formula gives the elements with the smallest numbers possible. The molecular formula gives the exact and proper formula, and it is derived from the empirical formula. If we are interested how much of the reactants are used ups and how much of the products that are formed, then this can be calculated based on mole ratios. It is the limiting reactant or limiting reagent that is taken up first, and this tells us how much product is formed. The theoretical yield gives us the maximum amount that can be produced or made. The actual yield is the yield given in an experiment. The actual yield is always smaller than the theoretical yield. Taking the ratio of the actual yield over theoretical yield and multiplying by 100% gives the percent yield.

Chapter 5
Various Reactions

Advance of Chemical Elements

Chemical changes involve the shuffling of atoms. A chemical equation shows the shuffling of the atoms on paper, a blackboard, a marker board, and even on the computer. They are represented to show the reactants and the products. Reactants are considered to be the starting materials and are represented on the left side by the symbol of the arrow. Products are the substances that are produced on the right by the symbol of the arrow. Bonds break and bonds form in a chemical reaction. According to the Law of Conservation of Mass, atoms cannot be created or destroyed. All of the atoms have to be taken account into. There needs to be the same number on both sides of the arrow. One needs to be able to balance a chemical equation.

The following reaction is an unbalanced equation $H_2 + O_2 \longrightarrow H_2O$. We notice that there are two atoms of hydrogen on the left hand side of the arrow, and there are two atoms of hydrogen on the right hand side of the arrow. At this point, the number of hydrogen atoms is balanced. However, there are two oxygen atoms on the left hand side of the arrow, and only one oxygen atom on the right hand side of the arrow. Placing a coefficient of two on the product side gives $2H_2 + O_2 \longrightarrow 2H_2O$. Now, the hydrogen atoms need to be balanced. A coefficient of two in front of H_2 gives

$$2H_2 + O_2 \longrightarrow 2H_2O$$

The equation now has the proper number of hydrogen and oxygen atoms on the left hand and on the right hand side of the arrow. The chemical equation is now balanced.

The identity of the reactants and products are determined by experiment. It is also important to know the physical states of the reactants and products. The symbol for a solid is (s), for a liquid is (l), for a gas is (g), and aqueous solutions are represented as (aq). The abbreviation lets us see and understand the physical state for each substance. The symbols are placed after the identity of the reactants and products. It is also important to keep in mind that when balancing equations, the identities of the molecules should never be changed. The formula for each molecule in the reactant and product should be the same. Atoms cannot be added or subtracted and subscripts cannot be changed.

When balancing a chemical equation, start with the most complicated molecule. Start balancing the carbon atoms, then the hydrogen atoms, and finally the oxygen atoms. It is best to balance diatomic molecules last since fractions can be placed in front of diatomic molecules. If a fraction is placed in front of a diatomic molecule and all the atoms are balanced, then the fraction can be eliminated by multiplying a number times the specific fraction to equal the numerical value one.

Chemical reactions can be considered as chemical changes. Examples of chemical changes include steel rusting, clothing being bleached, mixing ingredients, or fireworks. The possibilities of chemical reactions are endless. But there is more to it than that. We know a chemical reaction has occurred if there is a change in color, a solid forms, the presence of bubbles, or if there is a change in temperature based on the production or absorption of heat. Whenever there is a chemical reaction, there is always a chemical change. Atoms are being rearranged into atoms of new elements, compounds, or molecules. We represent chemical reactions as writing the starting materials or reactants on the left side of the arrow and the finished materials or products on the right hand side of the equation. An arrow or double arrow is used between the reactants and products to show whether the reaction took place in the forward direction or if the reaction is reversible.

Each of the elements, compounds, or molecules can be represented as a physical state. The solid state is represented as (s). The liquid state is represented as (l). The gaseous state is represented as (g). Finally, the aqueous state is represented as (aq). The aqueous state can also be thought of as a substance being dissolved in water. Consider the chemical equation below

$$CH_4(g) + 2O_2(g) \longrightarrow CO_2(g) + 2H_2O(g)$$

The table shows the number of atoms for each reactant and product. You notice there is a row for each of the elements carbon, oxygen, and hydrogen. The order of the elements does not matter when balancing the final chemical equation.

Table 5-1: Balancing Elements for the Reaction Above

Elements	Reactants	Products
Carbon	1	1
Hydrogen	4	4
Oxygen	1	1

You notice that the number of reactants and products are the same. This is always true because of the Law of Conservation of Mass. Atoms are neither created or destroyed. The same amount of atoms are always the same in both sides for reactants and products.

When balancing a chemical reaction, the formulas of the compounds cannot be changed. The subscripts also cannot be changed. We also cannot add or subtract formulas from the equation. The only thing we can do is place whole number coefficients in front of the formulas. It is also good to double check our work and make sure the same number of atoms are on both sides of the equation. There are millions of reactions in everyday life. How exactly the atoms rearrange each other in the reactants to form the products depends on the reaction itself and under certain laboratory conditions. Most of the time it is not easily predictable to determine the final products. A lot of these reactions are done by experiment to see what the final product(s) are. One can be sure that no matter what reaction takes place there will always be a balance of elements on both sides of the equation.

The driving forces of chemical reactions include a solid being formed, water being formed, a gas being formed, and the transfer of electrons. These are the predominant methods to get reactants to form products. But how exactly does a solid form for example when two different compounds are mixed with each other. This is where we now talk about solubility. Solubility has to do with the ability of elements, compounds, or molecules mixed with each other or not. Compounds such as nitrate salts are soluble. Salts of cations such as sodium and ammonium and anions such as chlorides are soluble. Exceptions include silver chloride and lead(II) chloride. Sulfate salts are also considered to be soluble. Exceptions include barium sulfate and lead(II) sulfate. Most compounds containing hydroxides are insoluble. Exceptions include sodium hydroxide and potassium hydroxide which are soluble in aqueous solution. Finally, anions such as sulfides, carbonates, and phosphates are insoluble.

Solubility is important to know because we would like to determine if a compound such as sodium chloride is soluble in water and is also a strong electrolyte. When sodium chloride dissolves in water, the compound breaks apart into its ions sodium and chloride. The sodium cation has a positive one charge, and the chloride anion has a negative one charge. Since they both each have charges, they can conduct electricity when an electrical conductivity apparatus is applied to the solution. Therefore, sodium chloride dissolved in water is considered a strong electrolyte because it conducts electricity. However, weak electrolytes such as acetic acid barely conducts electricity in water because the molecule itself does not 100% break apart into it cation and anion. If we were to put sugar into a beaker of water connected to an electrical conductivity apparatus we find the solution itself does not conduct electricity. Sugar does not break apart into ions when mixed with water.

We can describe the mixing process by writing its molecular equation, complete ionic equation, and net ionic equation. Consider the example when we mix sodium chloride and silver nitrate together in the laboratory which is a very common experiment. The products become solid silver chloride and sodium nitrate. This is called a precipitation reaction. The molecular equation is written as

$$NaCl(aq) + AgNO_3(aq) \longrightarrow AgCl(s) + NaNO_3(aq)$$

The complete ionic equation is considered as follows

$$Na^+(aq) + Cl^-(aq) + Ag^+(aq) + NO_3^-(aq) \longrightarrow AgCl(s) + Na^+(aq) + NO_3^-(aq)$$

The net ionic equation is written where it shows the main species in solution reacting together to form products. The ions that are the same on both sides of the equation are called spectator ions. Spectator ions don't react with each other to form the main products. On paper, we cancel the same species on both sides of the equation. In this case, the spectator ions in this example are sodium and nitrate. Therefore, the net ionic equation is

$$Ag^+(aq) + Cl^-(aq) \longrightarrow AgCl(s)$$

What would happen if we mixed a magnesium metal with gaseous oxygen? The product would be magnesium oxide. The reaction can be written as

$$2Mg(s) + O_2(g) \longrightarrow 2MgO(s)$$

This is an example of an oxidation-reduction or redox reaction. If we were to take a look at half reactions we would see that

$$Mg(s) \longrightarrow Mg^{2+}(aq) + 2e^-$$

$$O + 2e^- \longrightarrow O^{2-}$$

What exactly is happening here between theses elements? The magnesium metal loses two electrons and becomes oxidized. The magnesium cation therefore becomes smaller. Oxygen, on the other hand, gains two electrons and becomes reduced. The oxygen anion therefore becomes bigger. This explains why we call this an oxidation-reduction reaction.

If we mix an acid and a base together, we will always get salt and water. An example of this reaction is

$$HCl(aq) + KOH(aq) \longrightarrow H_2O(l) + KCl(aq)$$

This is called an acid-base reaction. The acid in this case is hydrochloric acid (HCl) and the base is potassium hydroxide (KOH). The products are liquid water H_2O and the salt is potassium chloride (KCl). There are ways to further classify redox reactions. An example is the combustion reaction. Combustion reactions are also known as redox reactions. They provide heat and electricity for our communities. It also provides a source of energy for transportation. An example of the reaction is

$$CH_4(g) + 2O_2(g) \longrightarrow CO_2(g) + 2H_2O(g)$$

Combination reactions are another type of redox reactions. The combining of elements, compounds, or molecules to form plastics and medicinal drugs has greatly enriched our lives. The formation of table salt is an example of a combination reaction. The reaction is as follows

$$2Na(s) + Cl_2(g) \longrightarrow 2NaCl(s)$$

Decomposition reactions are also redox reactions. They are useful for generating specific elements or substances that can be used for further reactions and that can be used for the production of new compounds or molecules. An example of the reaction is

$$2H_2O(l) \longrightarrow 2H_2(g) + O_2(g)$$

Balancing Redox Equations

A technique used in balancing redox equations is called the half-reaction method. One needs to split the reaction apart into two half-reactions. One of them is going to involve oxidation and the other one is going to involve reduction. Balancing equations such as these depend on if the reaction takes place in acidic or basic solution. Let us take a look at acidic solutions involving the half-reaction method. The first thing to do is to separate the equations. One of the half reactions should be an oxidation and the other should be a reduction. All of the elements should be balanced. The elements hydrogen and oxygen should be left alone. To balance oxygen, one adds water. To balance hydrogen, one adds H^+. The charges are balanced using electrons. Make sure the electrons are the same by multiplying the appropriate number of electrons to one or both half reactions. The half reactions should be added, and the species that are the same should be cancelled. The elements and charges should be balanced.

Let us also take a look at basic solutions involving the half reaction method. Split the reactions into an oxidation and a reduction reaction. All of the elements should be balanced. The elements hydrogen and oxygen should be left alone. To balance oxygen, one adds water. To balance hydrogen, one adds H^+. Hydroxide ions are added to equal the number of H^+ ions. The ions H^+ and OH^- combine to form H_2O. Eliminate the water molecules by subtraction. Water molecules are going to be on both sides of the equation. The last thing to do is to make sure the elements and charges are balanced.

Equations Describing Reactions

Equations can be used to describe reactions in solution. Let us consider what happens when we mix an aqueous solution of sodium sulfate with an aqueous solution of lead(II) nitrate. The resulting compounds lead(II) sulfate and sodium nitrate form. According to solubility rules and by experiment, lead(II) sulfate is the precipitate and sodium nitrate remains in aqueous solution. The molecular equation for this reaction is

$$Na_2SO_4(aq) + Pb(NO_3)_2(aq) \longrightarrow PbSO_4(s) + 2NaNO_3(aq)$$

Let us take a look at what happens in solution. The complete ionic equation is as follows

$$2Na^+(aq) + SO_4^{2-}(aq) + Pb^{2+}(aq) + 2NO_3^-(aq) \longrightarrow PbSO_4(s) + 2Na^+(aq) + 2NO_3^-(aq)$$

You notice here the strong electrolytes are shown as ions. Spectator ions are ions that don't take place in reactions such as the sodium cation and the nitrate anion in this reaction. The lead cation and the sulfate anion take place in the reaction. They form the precipitate lead(II) sulfate. Let us take a look at the reaction

$$Pb^{2+}(aq) + SO_4^{2-}(aq) \longrightarrow PbSO_4(s)$$

We call this the net ionic equation. These are the species that actually take part of the reaction.

For precipitation reactions, it is not easy to determine the kind of reaction that will happen when solutions are mixed together. We need to know the species that are present in the solution. Knowing the volume of the solution and the known molarity, we need to know the moles of the reactants in solution. For acid-base reactions, we need to know and concern ourselves with the species present in the solution when it is mixed. We know that an acid donates a proton, and a base accepts a proton. Acid-base reactions are also called neutralization reactions. If exact amounts react with each other, then the solution has been neutralized. For oxidation-reduction reactions, one or more electrons being transferred from one species to another is called an oxidation-reduction reaction or redox reaction. Examples of redox reactions are photosynthesis, combustion, and metabolic pathways. Oxidation means that there is an increase in oxidation state or there is a loss of electrons. Reduction means that there is a decrease in oxidation state or there is a gain of electrons. The oxidizing agent is the electron acceptor, and the reducing agent is the electron donor. In other words, the oxidizing agent is reduced and the reducing agent is oxidized.

Reactions involving Radiation

The chemistry of an atom is based on the number and arrangement of the electrons. The nucleus provides the positive charge that holds the electrons in atoms and molecules. The nucleus is very small in size, has a large density, and has a lot of energy that holds it together. The nucleus consist of neutrons and protons. Protons and neutrons are composed of smaller particles called quarks. We can think of the nucleus as being a collection of nucleons which are protons and neutrons. The number of protons in a nucleus is the atomic number, and it is represented by the symbol Z. The sum of the protons in a nucleus is called the mass number, and it is represented by the symbol A. Isotopes have identical atomic numbers, but their mass numbers are different. A nuclide refers to a particular member of a group of isotopes.

When we talk about thermodynamic stability, we refer to the potential energy of a nucleus in comparison to the sum of the potential energies of the component protons and neutrons. Kinetic stability refers to the probability a nucleus decomposes to a different nucleus. This is referred to as radioactive decay. Nuclei can be radioactive. The ejection of an electron is considered to be a beta particle. Stable nuclides are referred to as the zone of stability. Nuclides that have 84 or more protons are not stable in respect to radioactive decay. Spontaneous beta production occurs when there are too many neutrons. Spontaneous positron production occurs when there are too many protons. Light nuclides are stable when the number of protons equals the number of neutrons. In other words, this is when the neutron to proton ratio is one. Heavier elements have neutron to proton ratios greater than one that is required for nuclear stability as protons increase. Nuclides that have an even number of protons and neutrons are more stable than those with odd numbers. Protons or neutrons with numbers 2, 8, 20, 28, 50, 82, and 126 produce stable nuclides. This is similar to that in atoms where the number of electrons have 2, 10, 18, 36, 54, and 86 for the noble gases which have chemical stability.

Energy changes for the thermodynamic stability of a nucleus can be calculated based on Einstein's theory of relativity equation

$$\Delta E = \Delta mc^2$$

This is where c is the speed of light, Δm is the change in mass or mass defect, and ΔE is the energy change. A system can gain or lose energy which also gains or loses a quantity of mass. The mass of a nucleus should be less than the component nucleons. The process therefore is exothermic.

Energy changes for nuclear processes are very large compared to chemical reactions. Nuclear processes contain energy that could be harnessed and used. Thermodynamic stability for a nucleus is represented as energy released per nucleon. If the sign of the change in energy is negative then the process is exothermic. If the change in energy is positive, then the process is endothermic, and we can consider this as the binding energy of the nucleus.

Radioactive Decay

Radioactive nuclei decompose different ways. Either the mass number changes or does not change for the decaying nucleus. An alpha particle is considered to be the helium nucleus. Spontaneous fission is a decay process where the mass number of the nucleus changes. A heavy nuclide splits into two lighter nuclides that have similar mass numbers. Beta-particle production is a decay process where the mass number of the decaying nucleus stays constant. A beta particle has a mass number of zero. The atomic number for the beta particle is -1. In effect, the new nuclide is greater than one than the original nuclide. The overall effect producing a beta particle is to change a neutron to a proton. The beta particle is considered to be an electron. The nucleus does not emit electrons. The nuclide that is not stable makes an electron while it releases energy in the decay process.

A gamma ray is considered to be a high-energy photon. Producing gamma rays also makes nuclear decays and other particles in production. Emitting gamma rays is a way the nucleus with excess energy releases back to the ground state. A positron is considered to be a particle that has the same mass as an electron but the charge is positive. It is also considered to be an antiparticle. The overall effect is to change a proton to a neutron. The product nuclide has a higher neutron to proton ratio than the previous nuclide. Combining a positron and an electron together forms high-energy photons. Annihilation takes places. Forms of matter are interchanged. Electron capture takes place when an inner orbital electron gets captured by the nucleus. Gamma rays get produced with electron capture to give off excess energy. Sometimes, a radioactive nucleus does not go through one stable state. It can go through a decay process. A decay series is formed until a stable nuclide gets formed. In fact, rate of decay is formed by the negative change in nuclides per unit time which is also proportional to the number of nuclides. Half-lives for radioactive samples can also be calculated. Half-life is the time required for the nuclides to reach half its value.

Other processes of radioactivity include nuclear fusion and nuclear fission. Nuclear fusion involves two light nuclei combining together to form a heavier and more stable nucleus. Nuclear fission involves splitting a heavy nucleus into two smaller nuclei which have smaller mass numbers. Units of radiation include rads and rem. Radiation doses are measured in rads. The abbreviation rads stands for radiation absorbed dose.

The equivalence statement is 1 rad = 10^{-2} J. Energy dose of the radiation and the effects of causing biologic damage can be calculated based on rem. The abbreviation rem stands for roentgen equivalent for man. It is calculated as follows

Number of rems = (number of rads) x RBE

The abbreviation RBE represents how effective the radiation causes biological damage.

<u>Conclusion</u>

Reactions occur in aqueous solution where ions are hydrated. Strong electrolytes dissociate completely in solution and conduct an electrical current. Weak electrolytes barely dissociate into its ions and produce a small current. Nonelectrolytes do not conduct a current because they do not dissolve in water. Acids produce protons in solution. Bases produce hydroxides in solution. Acids are also considered to be proton donors, and bases are also considered to be proton acceptors. Molarity is one method to describe the concentration of a solution. We can define molarity as the moles of solute divided by the liters in solution. Standard solutions tell us that the molarity is known. Adding solvent to a solution means we are diluting the solution. When we are dealing with dilution, it is important to note that the moles of solute after dilution is equal to the moles of solute before dilution.

Molecular equations, complete ionic equations, and net ionic equations describe reactions in solution. Redox reactions have to do with electrons being transferred. Oxidation means that there is an increase in oxidation state and that there is a loss of electrons. Reduction means that there is a decrease in oxidation state and that there is a gain of electrons. Oxidizing agents accept electrons. Reducing agents donate electrons. The half reaction method is a technique for balancing redox equations. One of the half reactions is an oxidation, and the other half is a reduction. Balancing redox reactions depend on whether the reaction is in acidic or basic solution.

The stability of a nucleus has to do with the number of neutrons and protons. Some nuclides are stable and are found in the zone of nuclear stability. Unstable nuclei decompose into more stable nuclei by a process called radioactive decay. Alpha particles, beta particles, positrons, and/or gamma rays can be produced. Gamma rays are high energy photons. Radioactive decay processes follows kinetics, and half-life for a radioactive sample is the time for the amount of nuclide to reach half of the original amount. Thermodynamic stability of a nucleus can be compared to the energy of the nucleus to its component nucleons. A system can gain or lose energy which also results either in a gain or loss of a quantity of mass. The equation used to calculate the change in energy is Einstein's Theory of Relativity. The mass defect is the change in mass that can be obtained by comparing the mass for the nucleus to the sum of the component nucleons. It can also be used to calculate energy formed from the component nucleons. Binding energy is the amount of energy to decompose the nucleus to both protons and neutrons. Two light nuclei combine to form a more stable nucleus is called nuclear fusion. The splitting of a nucleus into lighter nuclei is called nuclear fission.

Chapter 6
Gas Laws

Bubbling

Matter has four physical states. These states are gas, liquid, solid, and plasma. We are interested in the property of gases in this section. Gases have been measured, and laws have been developed. Gases also behave in a certain way. The behavior of these particles tells us a lot about the properties of a gas. Atmospheric chemistry and the greenhouse effect are examples where gases take place. Gases can fill any type of container. They can be compressed easily, and they can mix with any other gas. Gases exert pressure on the surroundings. Gases also form the Earth's atmosphere. A barometer is used to measure atmospheric pressure. Atmospheric pressure and altitude can vary with each other. Units of pressure come from the height of a column of mercury. This depends on the pressure of the gas. The unit mm Hg is sometimes considered as the torr. The standard atmosphere is also allowed as a unit for pressure. It is abbreviated as atm. The equivalence statement can be represented as

$$1 \text{ atm} = 760 \text{ mm Hg} = 760 \text{ torr}$$

We can also define pressure as force per area. The formula for pressure is Pressure = Force/Area. The equation can be symbolized as

$$P = F/A$$

A Newton is the unit of force, and it is symbolized as N. The unit of area is meters squared. It has units N/m^2. This is also known as the Pascal. An equivalence statement can be represented by

$$1 \text{ atm} = 101,325 \text{ Pa}$$

Boyle's Law

Boyle was an Irish chemist who studied the relationship between pressure and volume. He used a J shaped tube to study how pressure and volume relate to each other. He multiplied the values together to obtain product values of pressure times volume. He found that the values were constant. This can be represented as

$$PV = k$$

This is called Boyle's Law. The letter P stands for pressure, V stands for volume, and k is a constant for a sample of air at a temperature. Boyle's Law can be represented graphically as $PV = k$ or $V = k(1/P)$. Note that P and V vary inversely with each other. Boyle's Law give good results at very low pressures. If a gas behaves like Boyle's Law, we can think of it as an ideal gas. Initial and final pressures and volumes can be represented by the equation where i stands for initial and f stands for final

$$P_i V_i = P_f V_f$$

Charles's Law

Jacques Charles was a French physicist who filled a balloon with hydrogen gas. He made the first flight using a balloon. He found that volume increases with temperature in a linear fashion. Volumes of gases go to zero at the temperature of -273 degrees Celsius. The Kelvin temperature corresponds this to be 0 K. The relationship between Kelvin and Celsius corresponds to

$$T_K = T_{oC} + 273$$

Volume and temperature are directly proportional to each other. This can be represented by the equation Charles's Law which is

$$V = bT$$

The value 0 K is known as absolute zero because a gas can't have a negative volume. The value 0 K has not been obtained, but temperature values such as 0.000001 K have been obtained in chemistry laboratories. Initial and final temperatures and volumes can be calculated from the following equation where i stands for initial and f stands for final

$$V_i/T_i = V_f/T_f$$

Avogadro's Law

Avogadro was an Italian chemist who came up with the idea that equal volumes contain the same number of moles. This is assuming that pressure and temperature remain constant. This is referred to as Avogadro's Law. Avogadro's Law is stated in equation form as

$$V = an$$

The letter V stands for volume, n is the number of moles, and a is considered to be a proportionality constant. Avogadro's Law works for gases at low pressures. Initial and final volumes and moles can be calculated from the following equation where i stands for initial and f stands for final

$$V_i/n_i = V_f/n_f$$

Ideal Gas Law

The ideal gas law is based on the following proportion. Volume is indirectly proportional to pressure. Volume is directly proportional to temperature. Volume is also proportional to the number of moles. We have the following relationships: $PV = k$, $V = bT$, and $V = an$ where k, b, and a are constants. These three equations can be combined to form what is called the idea gas law. The ideal gas law is

$$PV = nRT$$

The value R is known as the universal gas constant. The units can be expressed as anything it wants to be, however, the numerical value for R would be different. If we express the pressure in atmospheres, the volume in liters, the temperature in Kelvins, and the variable n as the number of moles, then experimentally we have the value 0.08206 L atm/mol K. The ideal gas law is also known as the equation of state. The state of the gas is a condition at a certain time. This is how we describe the state of a gas by knowing three of its properties. A gas behaves ideally if it obeys this equation. We can consider the ideal gas equation as a limiting law. It applies to low pressure and high temperatures. Pressure, volume, and moles can be expressed in any units as long as the units cancel with each other. Temperature always has to be converted to Kelvins. This involves adding the value 273.

Gas Stoichiometry

Suppose the gas contains 1.00 moles and is at 0 degrees Celsius and 1.00 atm. The volume of an ideal gas is calculated as follows

$$V = nRT/P$$

$$V = (1.00 \text{ mol})(0.08206 \text{ L atm/K mol})(273 \text{ K})/1.00 \text{ atm}$$

$$V = 22.4 \text{ L}$$

This value is considered to be the molar volume. The conditions 0 degrees Celsius and 1 atm is called standard temperature and pressure (STP).

Molar Mass of a Gas

Let us take a look at the relationship between gas density and molar mass

$$n = m/M$$

The value m is considered to be mass and M is the molar mass. Using the ideal gas equation and substituting in the expression above one obtains

$$P = nRT/V = (m/M)RT/V$$

$$P = mRT/MV$$

$$d = m/V$$

$$P = dRT/M$$

$$M = dRT/P$$

Dalton's Law of Partial Pressures

John Dalton in 1803 came up with his observations that the total pressure for a mixture of gases is the sum of the pressures for each individual gas. We can summarize his law of partial pressures as follows

$$P_{Total} = P_1 + P_2 + P_3 + P_4 + \ldots$$

The subscripts are for each individual gas. Each of the individual pressures are called the partial pressure for each individual gas if it were alone in a container. If the gas behaves ideally, the partial pressure can be used from the ideal gas law

$$P_1 = n_1RT/V \quad P_2 = n_2RT/V \quad P_3 = n_3RT/V \quad P_4 = n_4RT/V$$

In fact, we can calculate the total pressure of the mixture as follows

$$P_{Total} = P_1 + P_2 + P_3 + P_4 + \ldots$$

$$P_{Total} = n_1RT/V + n_2RT/V + n_3RT/V + n_4RT/V$$

$$= (n_1 + n_2 + n_3 + n_4)(RT/V)$$

$$= n_{Total}(RT/V)$$

The symbol n_{Total} represents the sum of the moles. The total number of moles is what we are interested in rather than the identity of the individual gas particles. Making these generalizations lets us know that the volume and force for each individual particle is not important.

Mole Fraction

The mole fraction is defined as the ratio of the number of moles to the total number of moles in a mixture. The symbol X is used to show the symbol for mole fraction. The symbol X which is called chi is a Greek lowercase letter. The equation for chi is

$$X_1 = n_1/n_{Total} = n_1 + n_2 + n_3 + n_4 + \ldots$$

Since $n = PV/RT$, then for each individual gas

$$n_1 = P_1V/RT \quad n_2 = P_2V/RT \quad n_3 = P_3V/RT \quad n_4 = P_4V/RT$$

The equation for mole fraction is and pressure is as follows

$$X_1 = P_1/P_{Total} \qquad X_2 = P_2/P_{Total} \qquad X_3 = P_3/P_{Total} \qquad X_4 = P_4/P_{Total}$$

Kinetic Molecular Theory of Gases

The Kinetic Molecular Theory is considered to be a model that explains how an ideal gas behaves. Let us look at some postulates on the kinetic molecular theory of gases. The first postulate states that the volume of each individual particle is assumed to be zero. They are small in comparison to the distance between each of them. The second postulate states that each particle is constantly in motion. The pressure of each gas depends on the collision of the wall of the container. The third postulate states that each particle has no force on each other. They don't attract or repel one another. The fourth postulate states that the average kinetic energy is directly proportional to the Kelvin temperature. Keep in mind that real gases do not follow these assumptions. Let us relate this model, the kinetic molecular theory model, to the ideal gas law. For Boyle's Law, if pressure increases then the volume of the container decreases. For Guy-Lussac's Law, if pressure increases then the temperature of the container with the molecules increases. For Charles's Law, if volume increases then the temperature of the container with the molecules inside it increases. For Avogadro's Law, if volume increases then the amount of moles in the container increases.

Daltons's Law: Mixture of Gases

The total pressure of a gas mixture os equal to the sum of the pressures of all the individual gases. According to the kinetic molecular theory, gas particles are independent and the volumes of each particle is not important.

Effusion and Diffusion

Graham's Law of Effusion

Rate of effusion for gas 1/Rate of effusion for gas 2 = $\sqrt{(M_2)}/\sqrt{(M_1)}$

Real Gases

$P(obs) = [nRT/V - nb] - [a(n/V)^2]$

where P(obs) is observed pressure

V is volume of the container

nb is volume correction

$a(n/V)^2$ is pressure correction

Van der Waals equation

$[P(obs) + a(n/V)^2] \times (V - nb) = nRT$

where $P(obs) + a(n/V)^2$ is the corrected pressure

$V - nb$ is the corrected volume

Chapter 7

Thermochemistry

Laboratory Glass

Energy is important in our daily lives. We eat food to give us energy to live. It is just as important for the consumption of coal and oil used for manufacturing and transportation. Fossil fuels have been used for these purposes. Petroleum is used a lot everyday. Supplies of oil have been low these past few years. No one knows for sure about the needs for supplies in the future. We need to find alternative products of petroleum if we are to maintain our standard of living. Having an understanding of the chemistry of energy can help us live a better life if we are successful finding innovative ways to provide oil, coal, petroleum, and any other useful piece of matter.

Fossil fuels are important for our civilization, but they have also created a problem. Fossil fuels produce waste products that affect the environment. Fuel is made up of organic compounds called hydrocarbons that react with oxygen in the atmosphere to produce carbon dioxide and water vapor. The carbon dioxide produced gets consumed by plants through the process of photosynthesis. It can also be used to form carbonate materials found on Earth. However, the balance of carbon dioxide produced and received is not in equilibrium. Carbon dioxide in the atmosphere is increasing. Carbon dioxide in our atmosphere absorbs heat radiation from the surface of Earth, and re-radiates it back towards the Earth. This process is important in controlling the Earth's temperature. Unfortunately we have an increasing amount of carbon dioxide produced in our atmosphere that is warming the Earth which in turn is causing significant climate changes. There is also the problem that impurities in fossil fuels react with certain substances in air to cause air pollution. All of this mentioned is taking place in our macroscopic world. Energy is also important in the microscopic world. Living organisms need a balance of energy with its internal and external environment. A cell can be thought of as a small chemical factory that uses energy from different types of reactions. Cellular respiration extracts energy from sugar to carry out the functions of the cell. The processes of cells or engines using energy are governed by the same principles.

Energy

Defining energy precisely is not that easy. But we can say that energy is the capacity to do work or the ability to produce heat. Let us take a look at the heat transfers that goes along with chemical processes. What we should know about energy is that it is still conserved. The Law of Conservation of Energy states energy is converted from one form to another, but it can neither be created or destroyed. Energy in the universe is constant. We can classify energy either as potential energy or kinetic energy. Potential energy is energy due to position in space. Both attractive and repulsive forces of a magnet can lead to potential energy. Kinetic energy has to do with the motion of an object. It depends on both mass and velocity. The equation for kinetic energy is

$$K.E. = (1/2)mv^2$$

It is possible to convert energy from one form to another. This can be achieved by the movement of balls on a pool table. Balls standing still have potential energy, and if you hit a ball the kinetic energy of a ball gets converted to the kinetic energy of another ball. The final ball comes to a stop and now has potential energy.

The amount of energy that is required to raise the temperature of a gram of water by a degree Celsius is called a Calorie. Calories can be changed to units called joules. The equivalence statement between calories and joules is 1 cal = 4.184 J. The equivalence statement can be converted to ratios in order to convert one unit to another. The specific heat capacity is also considered to be the specific heat. It is the amount of energy that changes temperature of one gram by a degree of Celsius. The equation for energy is energy is equal to mass x specific heat capacity x (final temperature - initial temperature). It can be represented in equation form as

$$Q = ms\Delta T$$

This is where $\Delta T = T_f - T_i$ and the value Q represents energy.   Energy can have units of calories, kilocalories, joules, or kilojoules. The mass of a sample is measured in grams. It is represented by the small letter m. The specific heat capacity is measured in Joules per gram degree Celsius. It is represented as small letter s. The final temperature that is represented as T_f is in degrees Celsius. The initial temperature that is represented as T_i is also in degrees Celsius.

The terms heat and temperature are different. Temperature is thought of as the random motions of particles in a substance. Heat is the transfer of energy between any two objects due to a difference in temperatures. Let me clarify that heat is not part of a substance contained from the object. Work is considered to be force acting over distance. Work and heat are two ways to transfer energy. The change in energy does not depend on pathway. However, work and heat do depend on the pathway. A state function or state property depends on its current state. It does not depend how the system got there. Changes going from one state to another does not depend on the pathway between the two states. Energy is considered to be a state function. Work and heat are not state functions.

Chemical Energy

Combustion of methane heats our homes according to the following reaction

$$CH_4(g) + 2O_2(g) \longrightarrow CO_2(g) + 2H_2O(g) + energy$$

The universe is divided into parts. It consists of the system and its surroundings. The system is what we should focus on. The surroundings is everything else outside the system. The system consists of reactants and products of the reaction. The surroundings could be the center where the reaction is taking place. If heat is given off, it is said to be exothermic. Energy is flowing out of the system. On the other hand, reactions that absorb energy from the surroundings are considered to be endothermic. The energy released as heat comes from the difference in potential energies between products and reactants. The energy that is gained by the surroundings is equal to the energy lost from the system. The flow of heat to the surroundings gives a lower potential energy for the reacting system. During an exothermic reaction, the potential energy stored in the chemical bonds get converted to thermal energy by heat. The chemical bonds in the products are stronger than the reactants. More energy gets released by forming new bonds than that consumed to break the bonds in the reactants. The change in potential energy is transferred to the surroundings as heat.

Endothermic reactions happen in reverse. Energy flows into the system as heat. The potential energy of the system is increased. The products have a higher potential energy than the reactants. The bonds are also weaker at a higher potential energy. Thermodynamics is considered to be the study of energy and its interconversions. The First Law of Thermodynamics follows The Law of Conservation of Energy. It states the energy of the universe remains constant. The internal energy, represented as U, is defined as the sum of kinetic and potential energies in the system. The internal energy can be changed by heat, work, or perhaps both depending on the conditions. The equation

$$\Delta U = q + w$$

represents the equation for internal energy, q represents heat, and w represents work. Thermodynamic values have two parts which is a number and a sign. The number is the magnitude and the sign is the direction of the flow. The sign tells us the system's viewpoint. Energy flowing into the system is an endothermic process where q is equal to a positive value. This is where the system's energy is increasing. However, energy can flow for the system in an exothermic process. This is where q is equal to a negative value. This is when the system's energy is decreasing.

Likewise, a system doing work on the surroundings indicates the sign for w to be negative. This is when energy is flowing out of the system. However, the surroundings doing work on the system is when w is positive. In this case, energy is flowing into the system. Work done by a gas is considered as expansion. Work done to a gas is considered compression. Consider a gas inside a cylindrical container with a movable piston. The pressure of the gas is P = F/A where F is a force acting on the piston that has an area A. Work is force times distance. If the piston moves by a change in height, then the work equation becomes W = F x Δh. Combining both expressions gives us W = P x A x Δh. The volume of a cylinder is equal to the area of the piston multiplied by the height of the cylinder. The change in volume moving a distance Δh is ΔV = A x Δh. The work equation then becomes W = P x A x Δh = PdΔV. However, w and ΔV have opposite signs. The equation for work is

$$w = -P\Delta V$$

Enthalpy

The enthalpy of a system is defined as

$$H = U + PV$$

The value U is the internal energy, P is the pressure, and V is the volume. Internal energy, pressure, volume, and enthalpy are all state functions. Changes in enthalpy do not depend on the pathway between two states. At constant pressure, the change in enthalpy for a system equals energy flow as heat. Heat flow is a measure of a change in enthalpy for an ideal system. The heat of reaction and change in enthalpy mean the same thing. For chemical reactions, the change in enthalpy is given by the difference in enthalpies of products and reactants. If the products for a reaction have a greater enthalpy than reactants, the change in enthalpy is considered to be positive. Heat is absorbed by the system. The reaction is considered to be endothermic. However, if the enthalpy for the products is less than the reactants, the change in enthalpy is considered to be negative. Decreases in enthalpy can be achieved by heat being generated. The reaction is thus considered to be exothermic.

Calorimetry

A calorimeter is a device that is used experimentally to determine the heat with a chemical reaction. Calorimetry is the science of measuring heat. It is based on observing a change in temperature when a body absorbs or discharges energy as heat. Each substance responds differently when it is heated. One substance could require a lot of heat energy, while another substance could require a smaller amount of heat. Heat capacity measures this property. It is C = heat absorbed/increase in temperature If an element or a compound is heated, the amount of energy depends on the amount of substance present. If we are going to talk about the heat capacity, we need to know the amount of substance. The specific heat capacity is the heat capacity given per gram of substance. The units are $J/^{\circ}C*g$ or $J/K*g$. If the heat capacity is given per mole of substance, it is referred to as molar heat capacity. It has the units $J/^{\circ}C*mol$ or $J/K*mol$. Looking at measured specific heat capacities of substances, we would find it takes less energy to change the temperature of 1 gram of metal by 1 degree Celsius than 1 gram of water.

Let us look at constant-pressure calorimetry. We can set up the following experiment. Two nested styrofoam cups, a cover with a stirrer, and a thermometer in it. The outer cup provides extra insulation. The inner cup contains the solution where the reaction takes place. This is considered to be a constant-pressure process because atmospheric pressure is also constant. Constant-pressure calorimetry can be used to determine the changes in enthalpy or heats of reaction that takes place in solution. According to these conditions, a change in enthalpy equals heat. Suppose we mix equal amounts of hydrochloric acid and sodium hydroxide at room temperature in a calorimeter. As the reaction is mixed and stirred, there exists an increase in temperature. The net ionic equation is

$$H^+(aq) + OH^-(aq) \longrightarrow H_2O(l)$$

Since there is an increase in temperature, the chemical reaction itself is releasing energy as heat. The energy that is released increases the random motions of the molecules. This increases the temperature. The energy released is determined based on an increase in temperature, the solution's mass, and the specific heat capacity. Keep in mind we are treating the system under ideal conditions. The calorimeter should not absorb or leak any heat. The solution is treated as if it was pure water with a density of 1.0 g/ml. If two reactants are mixed, and the solution gets warmer, the reaction is considered to be exothermic. Endothermic reactions cool the system. It is also important to know the required heat to raise the temperature of water by 1 degree Celsius. If we look at the specific heat capacity of water, we find it to be 4.18 J/°C*g. It takes 4.18 J of energy to raise the temperature of 1 gram of water by 1 degree Celsius.

The change in enthalpy for a neutralization reaction is $Q = ms\Delta T$ which translates to Energy released = Energy absorbed = specific heat capacity x mass of solution x change in temperature. Heats of reactions are extensive properties. It depends on the amount of substance or reactants. An intensive property does not relate to the amount of substance. An example of an intensive property is temperature. Calorimetry experiments can also be carried out at constant volume. A "bomb" calorimeter is used to study reaction conditions at constant volume. Reactants are weighed and put inside a durable rigid steel container. It is then ignited. One can measure the energy change by measuring an increase in temperature of the water and other parts in the calorimeter. The change in volume is equal to zero for constant-volume processes. Work is also equal to zero based on the equation $w = - P\Delta V$. The following equation holds true $\Delta U = q_v$.

Hess's Law

Enthalpy is a state function. The change in enthalpy from an initial state to a final state does not depend on the pathway. When reactants get converted to products, the change in enthalpy is the same value whether the reaction occurs in one step or in a series of steps. This is called Hess's Law. The overall reaction can either be given as the value ΔH or summing up the values of ΔH's. If the direction of the reaction is reversed, the sign of ΔH is charged. The magnitude of ΔH is proportional to the reactants and products. Multiplying the coefficients in a reaction by an integer causes the ΔH value to be multiplied by that same integer. Remember the sign of ΔH tells us the direction of heat flowing at constant pressure. The magnitude of ΔH is proportional to the reactants and products because ΔH is an extensive property. It depends on the amount of substance that is reacting.

Standard Enthalpies of Formation

If a reaction is studied at constant pressure, the enthalpy change can be obtained using a calorimeter. The process, however, can be difficult to carry out. Some reactions won't work under these conditions. Converting graphite to diamond won't work. The process is too slow under normal conditions. However, the change in enthalpy can be calculated from heats of combustion and then change its sign. The standard enthalpy of formation (ΔH_f^o) is the change in enthalpy that forms one mole of a compound from its elements when all their substances are in their standard states. A degree symbol shown for a given thermodynamic function shows the corresponding process that is carried out under standard conditions. The standard state is a reference state. Thermodynamic functions depend on concentrations or pressures of substances involved. Since thermodynamic functions depend on concentrations or pressures, a reference state must be used to compare thermodynamic properties of two substances. Most thermodynamic properties we measure changes the property. However, we don't have a method for determining absolute values of enthalpy. Enthalpy changes can be measured by carrying out heat-flow experiments.

The following are conventions of standard states. The standard state of a gas has a pressure of 1 atmosphere. A substance in a condensed state whether it is a liquid or solid has the standard state of a pure liquid or solid. A substance in a solution has the standard state of a concentration of 1 M. The standard state of an element exists under conditions of 1 atm and 25 degrees Celsius in their specific form. Remember that enthalpy is a state function. Hess's Law can be used for a pathway from reactants to products and add the enthalpy changes along a pathway. Take the reactants apart into their elements in their standard states and form the products from these elements. The pathway works for any reaction since atoms are conserved in a chemical reaction. A change in enthalpy for a given reaction is calculated by subtracting the enthalpies of formation of reactants from enthalpies of formation of products. This can be represented as the following equation

$$\Delta H^{\circ}_{rxn} = \sum[n_{prod}\Delta H^{\circ}_{f}(products)] - \sum[n_{reac}\Delta H^{\circ}_{f}(reactants)]$$

where the $\sum$ means take the sum of the terms and n_{prod} and nr_{eac} represents the moles of each product or reactant. Elements are not part of the calculation because elements do not require a change in form. The enthalpy of formation for an element in the standard state is zero. This is the reference point for calculating changes in enthalpy of reactions.

Energy Resources

Plants, petroleum, coal, and natural gas provide us with a measure of energy that come from the sun. Plants store energy through the process of photosynthesis. The energy from plants can be obtained by burning the plants themselves or decay products being formed that has been converted to fossil fuels. Petroleum and natural gas were probably formed from the remains of marine organisms. Petroleum is considered to be a thick, dark liquid made up of hydrocarbons. Natural gas contains mainly methane as well as ethane, propane, and butane. If we are going to use petroleum efficiently, it must be separated into different fractions. This can be done by the process of boiling the petroleum mixture. Lighter molecules are considered to have the lowest boiling points. They are boiled off first. The heavier molecules are left behind since they have higher boiling points. Petroleum that contains about 5 to 10 carbons is used for gasoline. Kerosene Jet fuel contains 10 to 18 carbons. Diesel fuel, heating oil, and lubricating oil contain 15 to 25 carbons. Asphalt contains more than 25 carbons.

Pyrolytic cracking is a process where we can increase the yield of gasoline. Heavy molecules of the kerosene fraction are heated to high temperatures. The molecules break into smaller pieces of hydrocarbons in the gasoline fraction. Additions such as tetraethyl lead were added to gasoline as an "antiknock" agent for engines. Air pollution from automobile exhaust was another problem. Catalytic converters were then added to car exhaust systems. However, the converters were destroyed by lead.

Leaded gasoline also increased the amount of lead in the environment where it is unsafe for animals and humans. The use of lead in gasoline has been phased out.

Coal was made from plants that have been buried and put under high pressure and heat for a long time. Plants have a lot of cellulose. Its molar mass is about 500,000 g/mol. Plants and trees flourished at many places and times. Eventually, they died and were buried. Chemical changes reduced the amount of oxygen and hydrogen of these cellulose molecules. Coal goes through four stages which is lignite, sub bituminous, bituminous, and anthracite. Each stage increases and there is always a higher carbon-oxygen and carbon-hydrogen ratio. Coal has different types of composition that depend on its age and location. The more carbon available in coal, the greater its energy content. Anthracite is more valuable of a coal than lignite. Coal is important because it is used as fuel in the U.S.A. It gives about 23% of our energy. As long as the supply of petroleum goes down, energy from coal increases. There are some drawbacks for the use of coal. It is expensive and dangerous to obtain from underground. Coal that has a high content of sulfur gives air pollutants such as sulfur dioxide. The sulfur dioxide can then turn into acid rain. Coal also has carbon, and when it is burned it forms in air. Too much of it causes global warming and affects Earth's climate.

Earth takes in a lot of radiant energy from the sun. Approximately 30% gets reflected back into space. The remaining energy goes through the atmosphere and reaches Earth's surface. Plants absorb some of this energy for photosynthesis. Oceans also take into this amount of energy to evaporate small amounts of water. Most of it, however, is absorbed by soil, rocks, and water that increases the Earth's surface temperature. The energy is radiated from the heated surface as infrared radiation or heat radiation. Earth's atmosphere is transparent to visible light. However, it does not allow all of the infrared radiation to go back into space. Molecules such as H_2O and CO_2 absorb infrared radiation and then radiate it back to Earth. A net amount of thermal energy is retained by Earth's atmosphere. This causes the Earth to be a lot warmer than it is supposed to be. We can think of the atmosphere like a glass of a greenhouse. The glass of a greenhouse is transparent to visible light, but it absorbs infrared radiation. This causes the temperature of a building to be warm.

Earth's surface temperature is controlled by the amount of carbon dioxide and water content of the atmosphere. The water content of the atmosphere is controlled by the water cycle. Evaporation and precipitation occurs in the water cycle. Infrared radiation is blocked by CO_2, methane, and other greenhouse gases. Increasing the Earth's temperature by as much as 3 degree Celsius can cause great changes in climate. This would affect crops and our natural resources. Methane is also a greenhouse gas that is a lot more potent than carbon dioxide. This is found in countries with lots of animals. Methane is produced in animal's rumen. Australia, for example, has a lot of sheep and cattle that produce methane in the rumen. Efforts have been made to reduce and lower the number of archae in the digestive system.

<u>New Energy Sources</u>

Potential energy sources include the sun, nuclear processes, biomass, and synthetic fuels. Using coal a different way is one alternative source of energy. Making a gaseous fuel or substances that have high boiling points such as solids or thick liquids. Converting coal from a solid to a gas requires the size of the molecules to be reduced. Coal has to be broken down in a process called coal gasification. Coal is treated with oxygen and steam at high temperatures. Carbon-carbon bonds are broken. Carbon-hydrogen and carbon-oxygen bonds are formed as the fragments that react with water and oxygen. The product that is obtained is a mixture of carbon monoxide and hydrogen called synthetic gas or syngas and methane gas. Syngas can react with oxygen and release a lot of heat in a combustion reaction. This makes the gas a useful fuel. Syngas can also produce other fuels. Syngas can be converted to methanol. Methanol can also be converted directly to gasoline. Coal slurries can also be used as coal. Coal is pulverized and mixed with water to make a slurry.

Hydrogen can also be used as a fuel. Hydrogen can also react with oxygen to produce water. Burning methane, decomposing water by electrolysis, and thermally decomposing water are possibilities because it is expensive and the amount of products would be low. Costs, storage, and transport are still problems that need to be considered when it comes to using hydrogen. If we can overcome these problems, then it can be possible to use hydrogen as a fuel. The best use of hydrogen is to use metals that absorb it to form solid metal hydrides. Hydrogen gas would have to be pumped into a tank where the solid metal is in its powdered form. The powdered form would absorb hydrogen to form a hydride. The hydrogen can then be used for combustion in the engine. This releases H_2 from the hydride.

Hydrogen powered vehicles might be the future. Hydrogen in automobiles can be used to power fuel cells. Oil shale contains kerogen, carbon-based material from porous rock formations, could be used as a source of energy. Ethanol produced from fermentation or fuel-grade ethanol from corn have the potential to be used as fuel. Gasohol, which is an alcohol-gasoline mixture, have also been used in car engines as a fuel. Methanol could possibly be used as a fuel additive. Methanol could alleviate air quality problems and could possibly be a major source of energy. Seed oil squeezed from seeds could also be used in the future. Seed oil would be a good thing to use because it is renewable. Oil-seed plants that can survive soil and climatic conditions is something that can also help us use as a fuel for the future.

<u>Conclusion</u>

Energy is defined as the ability to do work and/or to produce heat. The First Law of Thermodynamics states that energy is conserved and can be converted from one form to another. It cannot be created or destroyed. Heat is a way to transfer energy. Chemical reactions can either give off or absorb heat. Reactions that give off heat are called exothermic reactions. Reactions that absorb heat from its surroundings are called endothermic. We can think of energy as a state function. This means energy does not depend on the pathway from one state to another. Calorimetry is used to measure heat flow for a reaction. This is done by measuring changes in temperature when a body absorbs or releases heat. Constant pressure processes involve heat flowing that is equal to a change in enthalpy. Enthalpy is a state function. Enthalpy changes going from reactants to products is the same whether the reaction takes place in one step or a series of steps such as Hess's Law. The enthalpy change can be calculated by adding up the change in enthalpy values of each step to give a net reaction. The enthalpy values can be calculated from the standard enthalpies of formation of reactants and products. Standard enthalpy of formation is an enthalpy change that involves forming one mole of a compound from its elements. The substances are supposed to be in their standard states. The combustion of fuels made from carbon produce carbon dioxide and water. Both of these molecules control Earth's surface temperature. Too much carbon dioxide in the atmosphere can produce global warming due to the greenhouse effect. Alternatives to fuel like syngas, hydrogen, and other carbon based complex mixtures could be used to replace petroleum and its products.

Chapter 8

Atomic Theory

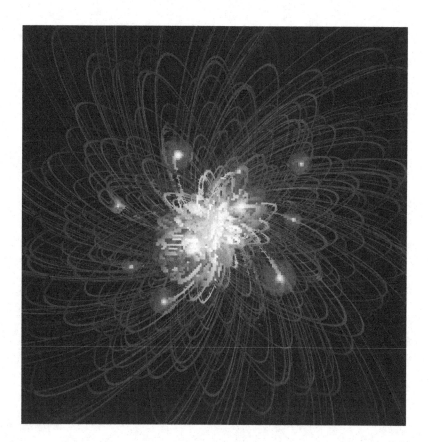

Electromagnetic Field-Nuclear Radioactive Core

Greek philosophers Democritus and Leucippus around 400 B.C. first suggested the idea of atoms based on what they know. For a long time, there was no experimental evidence to support that atoms do exist. Lavoisier and other scientists were the first to come up with scientific data based on quantitative measurements. Based on these experiments, John Dalton came up with the idea of the atomic theory. More studies have taken place to understand the internal structure of an atom. Each element contains a certain number of atoms. Many elements also show similar behavior based on chemical and physical properties. Many of these similarities led to the making of the Periodic Table of Elements. The modern atomic theory of atomic structure takes into account the periodicity of electron arrangement of atoms. Let us take a look at some important topics covered in physics that led up to the idea of atomic structure. Quantum mechanics was developed to explain the behavior of light and atoms. Let us first take a look at the properties of light. We call this electromagnetic radiation.

Electromagnetic Radiation

Energy goes through space by a process known as electromagnetic radiation. Examples of electromagnetic radiation include the light shining from the sun, heat from a fireplace, a microwave oven cooking food, and X-rays used by radiologists and dentists. All of these different types of radiant energy have wavelike behavior and travel at the speed of light. Waves can be described in three ways. Wavelength which is symbolized by the lower-case greek letter lambda which corresponds to the distance between two peaks or troughs in a wave. The frequency which is symbolized by nu is the number of cycles per second corresponding to a point in space. Speed is how fast the wave is traveling through space. There is an inverse relationship between wavelength and frequency. The equation lambda times nu which is equal to c can be used to calculate wavelength or frequency given either one of the conditions and the value of the speed of light. The speed of light is 2.9979×10^8 m/s and can simply be rounded to 3.0×10^8 m/s. The symbol lambda is the wavelength given normally in meters and the symbol nu is the frequency normally given in cycles per second. The unit per second is 1/s or s^{-1} which is considered to be Hertz. Radiation is a means of transferring energy. Electromagnetic radiation is ranked according to increasing wavelength and decreasing frequency as follows: Gamma rays, X-rays, Ultraviolet, Visible, Infrared, Microwaves, and then Radio Waves.

Matter

Matter consists of particles. Electromagnetic radiation is described as a wave. Particles have mass and its position in space can be described. A wave's position in space cannot be specified. Max Plank postulated energy has either energy gained or lost by whole-number multiples of the quantity hv. The value of h is experimentally determined to be 6.626×10^{-34} J * s. This is called Plank's constant. The change in energy is represented as $\Delta E = nhv$ where n is an integer, h is Plank's constant, and v is the frequency of electromagnetic radiation. Energy is actually quantized. Each small size of energy is called a quantum.

Albert Einstein came up with the idea that electromagnetic radiation is quantized. He suggested that electromagnetic radiation to be a stream of "particles" which he called photons. The energy of each photon is

$$E(photon) = hv = hc/\lambda$$

where h is Plank's constant, v is frequency, and λ is wavelength. Another equation he came up with was

$$E = mc^2$$

This equation tells us that energy has mass. Substituting the previous equation with this equation and solving for mass gives us

$$m = h/\lambda c$$

We can conclude that energy is quantized and occurs in discrete units called quanta. Electromagnetic radiation have wavelength and particle-like properties. This is referred to as the dual nature of light.

Louis de Broglie also made some contributions to our understanding of wave properties. The equation $m = h/\lambda c$ can be derived. If we use velocity instead of the speed of light, we have $m = h/\lambda v$. If we solve for λ, we have

$$\lambda = h/mv$$

This equation is known as the de Broglie equation. Diffraction is obtained when light is scattered from points or lines. Suppose X-rays are directed toward a crystal of sodium chloride, a diffraction pattern is obtained by showing bright spots and dark spots on the photographic plate. Scattered light either can interfere constructively or destructively. It interferes constructively to give a bright spot or it can interfere destructively to give a dark spot. Diffraction patterns are explained based on waves. Electromagnetic radiation has both wave-like and particle-like properties. Matter and energy are not that much different from each other after all. Energy is a form of matter, and matter shows these kind of properties. Matter exhibits particle and wave properties.

Hydrogen Atomic Spectrum

The emission of light by hydrogen atoms that are in their excited state was also studied. When a high energy spark is applied to a hydrogen gas, H_2 molecules tend to absorb energy. The single bonds containing H-H break apart. The hydrogen atoms tend to be in their excited state. The excess energy is released by emitting light of different wavelengths to produce an emission spectrum of a hydrogen atom. In fact, the hydrogen emission spectrum is called a line spectrum. It is called a line spectrum because when light from a hydrogen gas discharge tube passes through a slit and then to a prism, the detector detects different lines. These lines tell us that there are only certain energies that are allowed for the electron in a hydrogen atom. The energy for the electron is quantized. According to Plank's equation, changes in energy produce specific wavelengths of light that is emitted. Hydrogen's atomic spectrum tells us only certain energies are possible. Electron energy levels are quantized.

The Bohr Model

Niels Bohr came up with a quantum model for the hydrogen atom. He came up with the idea that an electron for hydrogen moves around the nucleus based on circular-type orbits. His model of the hydrogen atom energy levels was consistent with the hydrogen emission spectrum. The equation for the energy levels for an electron in a hydrogen atom is

$$E = -2.178 \times 10^{-18} \text{ J } (Z^2/n^2)$$

where n is an integer and Z is the nuclear charge. The negative sign for energy has some meaning to it. It means that the energy of the electron that is bound to the nucleus is lower if the same electron were at a far distance from the nucleus. This would be the case where the energy is zero and there is little or no interaction. The equation for energy then becomes E = 0. The equation can also be used to calculate a change in energy of an electron when the electron goes to a different orbit. The lowest energy state is referred to as the ground state, and the highest energy state is referred to as the excited state. The more negative energy means that the electron is more bound in that specific orbit. If the sign for a photon is produced when an electron jumps from a higher to a lower energy state, then the wavelength for a photon can be calculated as $\lambda = \Delta hc/E$. Energy levels that are calculated agree with values obtained from the hydrogen emission spectrum. However, Bohr's model does not work for any other atom except hydrogen. The truth is electrons do not move around the nucleus in orbits that are circular.

Quantum Mechanical Model of the Atom

Werner Heisenberg, Louis de Broglie, and Erin Schrodinger developed wave mechanics or quantum mechanics. Louis de Broglie came up with the idea that the electron which is a particle shows wave properties. Erwin Schrodinger studied the wave properties of an electron. The electron held by the nucleus can be thought of as a standing wave. Schrodinger developed a model for the hydrogen atom in which he assumed the electron to be a standing wave. Schrodinger's equation is in the form

$$H \psi = E \psi$$

where ψ is called the wave function. The wave function describes the coordinates of the position of an electron in space. The symbol H is considered to be the operator. The operator produces the total atom's energy applied to a wavefunction. The symbol E represents the total energy which is the sum of the kinetic and potential energies. Every solution of this equation contains a wave function ѱ that is based on the E value. Wave functions are normally called orbitals. Let us take a look at the quantum mechanical model of the atom. The lowest energy for a hydrogen atom is considered to be a 1 s orbital. We don't know how the electron is moving around the nucleus. The wave function does not tell us the pathway for the electron.

Let us take a look at the nature of an orbital. Werner Heisenberg came up with the idea that we cannot know the position and momentum of a particle at any given time. This is considered to be the Heisenberg Uncertainty Principle. Based on mathematics, the uncertainty principle is Δx times $\Delta(mv)$ greater than or equal to $h/4pi$ where the symbol Δx is considered to be the uncertainty of a particle's position, $\Delta(mv)$ is the uncertainty in a particle's momentum, and h is Plank's constant. The more accurate we try to determine the particle's position, the less accurate the momentum will be. The uncertainty principle when it is applied to the electron tells us that we do not know the motion of the electron when it is moving around the nucleus.

The Wave Function

Let us take a look at the square of the wave function. The square of the function tells us that the probability of finding an electron at some point in space. The relative probability of trying to find an electron is the ratio of each of the squares of the wave function is equal to the inverse of the positions. This quotient is the ratio of probabilities finding the electron at two different positions. It does not tell us when and how the electron will be between positions. This is consistent with the Heisenberg Uncertainty Principle. The square of the wave function tells us the probability distribution. The intensity of color tells us the probability value near a point in space. This is considered to be an electron density or electron probability. An electron density map of this type is called an atomic orbital. The probability of finding an electron is when it is closest to the nucleus. It is less probable of finding an electron when it increases from the distance from the nucleus. The quantum mechanical model tells us that the motion of an electron is not known. This is a probable distance where the electron is to be found. The size of an orbital can't be defined precisely. When we describe the size of a hydrogen 1 s orbital, we look at the radius of the sphere that surrounds 90% of the total electron probability. About 90% of the time an electron is found inside a sphere. All of this information describes the lowest-energy wave function for a 1 s orbital for a hydrogen atom. Orbitals are considered to be wave functions. They are usually thought of as three dimensional density maps.

Quantum Numbers

Solving the Schrodinger equation for a hydrogen atom gives us many wave functions or orbitals. The orbitals are characterized by quantum numbers. The principle quantum number is symbolized by the letter n. It has values 1, 2, 3….. It has to do with the size and energy of the orbital. Increasing the value of n gives us higher energy. The electron is not that bound to the nucleus. The angular momentum quantum number is symbolized by the value of l. It has values ranging from 0 to n - 1.

This has to do with the shape of an atomic orbital. Values of l for a particular orbital is given a letter. The value l = 0 is s, l = 1 is p, and l = 3 is f. The magnetic quantum number is symbolized by m_l. It has values between l and -l including 0. It has to do with the orientation of the orbital. Each set of orbitals or subshell is given by the value of n and the letter for l. The number of orbitals per subshell is s = 1, p = 3, d = 5, and f = 7.

Shapes, Energies and Spin States

An orbital can be thought of as a probability distribution. An orbital can also be thought of by the surface that surrounds about 90% of total electron probability. The shapes of each of the s orbitals are different. Nodes or nodal surfaces are areas of zero probability. As the value of n increases, the number of nodes increases. The number of nodes for s orbitals is n - 1. The p orbitals have two lobes that are separated by a node at the nucleus. The 3p orbitals have the same surface shapes, but they are bigger. As the value of n increases, the surface gets bigger. The d orbitals begin with the value n = 3. There are five 3d orbitals. Four of the orbitals look like cloverleaf shape. The fifth orbital has two lobes and a belt that is centered in the plane. The f orbitals have more complex shapes than the d orbitals. For a hydrogen atom, the energy is based on its value of n. Degenerate orbitals have the same energy with the value of n being the same. Hydrogen has only one electron, and it can occupy any of its orbitals. The electron is normally found in its ground state which is in the 1 s orbital. The ground state is also called the lowest energy state. Adding energy to an atom causes the electron to go to a higher-energy orbital. The higher energy state is considered to be the excited state.

Electrons have a magnetic moment with different possible orientations in the presence of an external magnetic field. The electron spin quantum number is symbolized as m_s. It can have one of the values as +1/2 and -1/2. The electron can spin in opposite directions. The Pauli Exclusion Principle states that electrons no more than two of them will have the same set of four quantum numbers. The quantum numbers are n, l, m_l, and m_s. Different values of m_s must be specified given the values of n, l, and m_l. Only two different values of m_s are specified. An orbital can only hold two electrons, and they have to be opposite spins.

Polyelectronic Atoms

Polyelectronic atoms are atoms that have more than 1 electron. The electron correlation problem has to do with unknown electron pathways and electron repulsions cannot be exactly calculated. All polyelectronic atoms have the electron correlation problem. We need to make approximations. Each electron that is moving in a field of change is the result of a nuclear attraction and repulsion of all other electrons. An outer electron feels an attraction to a nucleus, but it also feels the repulsion of other electrons. Therefore, the electron is not that bound and held tightly to the nucleus. This means the electron is screened or shielded from the positive charge of the nucleus by repulsion of other electrons. Polyelectronic atoms become hydrogen-like orbitals. They have the same shapes for hydrogen; however, the sizes and energies are not the same. Keep in mind that hydrogen-like orbitals are degenerate or have the same energy. Polyelectronic atoms have the energy levels as follows: $E_{ns} <$ $E_{np} < E_{nd} < E_{nf}$. Electrons are arranged in the order s, p, d, and then finally f.

If we take a look at the radial probability distributions of 2s and 2p orbitals, we find that the electron in a 2p orbital is closer and has a maximum probability than the 2s orbital. However, even though an electron in a 2s orbital spends time a little farther from the nucleus, it has a small but a very important and significant amount of time near the nucleus. The 2s electron is said to penetrate to the nucleus more than the 2p orbital. An electron in a 2s orbital is more strongly attracted than an electron in a 2p orbital. We can therefore conclude that the 2s orbital is lower in energy than any of the 2p orbitals in terms of a polyelectronic atom. The same principle applies to the d orbitals. The radial probability distributions for the orbitals given the value of n = 3 is a bit more complex. The 3s electron is normally found far from the nucleus. A small and significant amount of time is found near the nucleus. We can say that the 3s electron penetrates a shield of inner electrons. The 3d orbital has the highest probability being near the nucleus than other 3s or 3p orbitals. However, since the 3d orbital is not near the nucleus, it is said to be of highest energy of the three given orbitals. The energies for the orbitals are $E_{3s} < E_{3p} < E_{3d}$. We can make a general rule and state that an orbital which allows its electron to penetrate the electrons and which are also shielded close to the nuclear charge causes the lowering of the energy of the orbital.

The Aufbau Principle

Let us take a look at the electron arrangements to account for the Periodic Table is organized. All of the atoms have the same kind of orbitals. Protons are added one by one to build up different kinds of elements. Elements are also added to hydrogen-like orbitals. This is referred to the Aufbau Principle. Hydrogen has one electron, and it is filled in the 1 s orbital. Helium has two electrons, and it is filled in the 1 s orbital. The configuration is $1s^2$ and the electrons have opposite spins based on the Pauli Exclusion Principle. Lithium has three electrons. It has the configuration $1s^2 2s^1$. Beryllium has four electrons. It has the configuration $1s^2 2s^2$. Moving over to the p block, the element boron has five electrons. It has the configuration $1s^2 2s^2 2p^1$. Carbon has six electrons. It has the configuration $1s^2 2s^2 2p^2$. Each electron is in a separate p orbital due to the repulsive nature of the electrons. Hund's rule states that there should be a maximum unpair of electrons in degenerate orbitals. The unpaired electrons have parallel spins. The configuration for nitrogen has seven electrons, and it is filled as $1s^2 2s^2 2p^3$. The configuration for oxygen which has eight electrons has the form $1s^2 2s^2 2p^4$. One of the orbitals has two electrons with opposite spins in one of the p orbitals. Likewise, the electron configuration for fluorine is $1s^2 2s^2 2p^5$, and the electron configuration of neon is $1s^2 2s^2 2p^6$.

Sodium has ten inner electrons, and it occupies the 1s, 2s, and 2p orbitals. The eleventh electron is the outer electron, and it occupies a 3s orbital. Sodium has the electron configuration $1s^2 2s^2 2p^6 3s^1$. The inner electrons are $1s^2 2s^2 2p^6$ and we can symbolize this as one of the previous noble gas elements [Ne]. We can also write the electron configuration for sodium as $[Ne]3s^1$. Magnesium has the configuration $1s^2 2s^2 2p^6 3s^2$ or $[Ne]3s^2$. Valence electrons are the outermost electrons for a given atom. Valence electrons are the electrons that are involved in bonding. The core electrons are the inner electrons. Elements in the same group contain valence electron configurations that are the same. The elements have similar chemical behavior. Potassium has the electron configuration $1s^2 2s^2 2p^6 3s^2 3p^6 4s^1$ or $[Ar]4s^1$. Calcium has the electron configuration $1s^2 2s^2 2p^6 3s^2 3p^6 4s^2$. The next elements scandium through zinc make up some of the d metals. The d block are called the Transition Metals. Scandium has the configuration $[Ar]4s^2 3d^1$, titanium is $[Ar]4s^2 3d^2$, and vanadium is $[Ar]4s^2 3d^3$. Chromium configuration is unusual. It has the configuration $[Ar]4s^1 3d^5$. These last two orbitals are half-filled and are stable. Manganese has the configuration $[Ar]4s^2 3d^5$, Co is $[Ar]4s^2 3d^7$, Fe is $[Ar]4s^2 3d^6$, and Ni is $[Ar]4s^2 3d^8$. The electron configuration for copper is $[Ar]4s^1 3d^{10}$. This is for copper where we have a half-filled orbital and a filled orbital that is more stable. Zinc has the configuration $[Ar]4s^2 3d^{10}$. The elements gallium through krypton involve filling the 4p orbitals.

Let us note some important points about the filling of orbitals. Orbitals that are (n + 1)s fill orbitals that are nd. The s orbitals are filled because of the penetration effect near the nucleus. This allows the s orbital to be lower in energy than the d orbitals. The lanthanide series has to do with the filling of the seven 4f orbitals. Sometimes, an electron can occupy a 5d orbital instead to it being filled in the 4f orbital. The energies are similar. The actinide series or the actinides has to do with the filling of the seven 5f orbitals. Sometimes, one or two can occupy a 6d orbital instead of a 5d orbital because the energies are similar. The main group elements are Groups 1A - 8A. These are called the representative elements. The group number corresponds to the total number of electrons in its valence shell. The quantum mechanical model is used to explain the arrangement of the Periodic Table of Elements. For a given group, similar chemistry takes place because of the same valence electron configurations. The principle quantum number is the only thing that changes when going down a group.

Periodic Trends

There are three types of periodic trends discussed in a first year chemistry course. The first trend is ionization energy. Ionization energy is the energy that is needed to remove an electron from an atom or ion in the gas phase. It is represented as $X(g) \longrightarrow X^+(g) + e^-$. Atoms or ions are in their ground state. The first ionization energy is labeled as I_1. It is the energy needed to remove the highest-energy electron from an atom. Second ionization energy is labeled I_2 etc. The first electron removed from a neutral atom or ion gives a positive charge. Increasing the positive charge makes the electron to bind more stronger. This allows the ionization energy to increase. Ionization energies can jump even higher if core electrons are to be removed. Core electrons are more attracted to the nucleus than valence electrons. As we move from left to right across a period, ionization energy increases. As we go from top to bottom, ionization energy also increases. Electrons being added in the same row don't completely shield an increasing nuclear charge when protons are added. Electrons are more strongly bound as we move across the periodic table. Electrons are being added to the same principle quantum number. Therefore, the values of the energies increase. First ionization energy tends to decrease when we go down a group. It decreases because electrons that are removed are farther away from the nucleus. As the principle quantum number increases, we find that the size of an orbital gets bigger. Electrons are easier to remove from bigger orbitals. There are exceptions to the trends in ionization energy. Ionization energy decreases going from Beryllium to Boron because electrons in the filled 2s orbital give shielding for electrons in the 2p orbital from the positive nuclear charge. Nitrogen to oxygen decreases in ionization energy due to extra electron repulsions in the 2p orbitals.

The second trend is electron affinity. Electron affinity has to do with energy change by adding an electron to an atom in the gas phase. It is represented by the equation

$$X(g) + e^- \longrightarrow X^-(g)$$

The more negative the energy value, a greater amount of quantity of energy is being released. Electron affinities tend to be more negative when we go from left to right across a period. The relationship between electron affinity and atomic number is explained by looking at the changes in electron repulsions when it comes to electron configurations. Going down a group, electron affinity tends to get more positive. The electron is added to distances that are increasing from the nucleus itself. Electron affinity values going down a group are small and there are a lot of exceptions. No clear explanation except attraction and repulsion of electrons have been thought of.

The third trend is atomic radius. Sizes of atoms can't be given exactly either. If we measure the distance between atoms, we can then obtain the atomic radii by taking half its value. If we are dealing with covalent compounds, then we can refer to the radii as covalent atomic radii. If we are dealing with metal atoms, then we can refer to the radii as metallic radii. Atomic radii increase from right to left and from top to bottom. Increasing effective nuclear charge also means shielding is decreasing when we go from left to right across a period. Valence electrons are pulled more closer to the nucleus. The size of the atom decreases. Atomic radius tends to increase when we go down a group. This happens because of the orbital sizes increase as the principle quantum level gets bigger.

Conclusion

Energy goes through space as electromagnetic radiation. We describe electromagnetic radiation by its wavelength, frequency, and speed. These can be related to each other by the equation lambda times nu which is equal to the speed of light. Max Plank came up with the idea that energy is either gained or lost based on quanta. These are based on whole-number multiples of hv. Einstein came up with the idea that electromagnetic radiation is a stream of particles called photons. The energy of the photon is hv. Electromagnetic radiation has both wave and particle properties. The hydrogen spectrum gives a line spectrum. Specific energies are permitted for an electron in a hydrogen atom. The Bohr Model for the hydrogen atom assumed the electron moved in circular orbits and a line spectrum was obtained when an electron moves from one orbit to another orbit. Although the Bohr Model was incorrect, it was a good starting point for the study of the quantum mechanical model.

The quantum mechanical model is based on an electron being described by a wave function. Wave functions are also referred to as orbitals. The Heisenberg Uncertainty Principle states that we cannot know both the position and momentum of a particle at the same time.

Squaring a wave function gives us the probability of finding an electron in space. Probability distributions are also called electron density maps. They give us the shapes of the orbitals. Orbitals can be described by the quantum numbers n, l, m_l, and m_s. The spin quantum number which is denoted m_s tells us the electron can spin in two different directions. The Pauli Exclusion Principle tells us that no more than two electrons are allowed to have the same four quantum numbers. The electron configuration of elements are described by protons being added one at a time to the nucleus to make new kinds of elements. Elements are also added to specific orbitals. We refer to this as the Aufbau Principle. The quantum mechanical model helps us explain periodic properties of elements based on the periodic table. Elements in the same group have similar valence electrons and similar chemical properties. Ionization energy, electron affinity, and atomic radius are used to describe periodic trends of the periodic table. They have to do with attractions, repulsions, penetration, and shielding.

Chapter 9
Concepts of Bonding

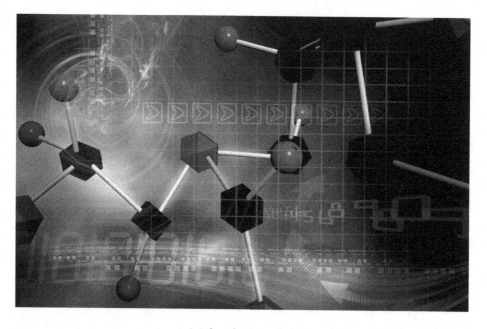

Molecule Samples

Let us take a look at the classes of compounds that describe different types of bonds. We will also take a look at models that describe structure and bonding. Chemical bonds can be defined as atoms being held together and functioning as a unit. There are many ways we can study materials. The melting point, boiling point, density, conductivity, solubility, charge distribution, and bond strength are characteristics of each materials that can be studied. When solid sodium chloride dissolves in water, the solution can conduct electricity. We know it conducts electricity because the sodium and chloride ions are in solution, and they are electrically charged. When the two elements react with each other, electrons from sodium are transferred to chlorine atoms to form ions. These ions can then come together to form solid sodium chloride. This happens because the system wants to be in the lowest possible energy. These oppositely charged ions attract each other to be in the lowest possible energy. The bonding forces have to do with the electrostatic attractions of oppositely charged ions. This is referred to as ionic bonding. The formation of an ionic bond or compound consists of a metal and a nonmetal. For a pair of ions, energy can be calculated based on Coulomb's Law. The equation for Coulomb's Law is the following

$$E = (2.31 \times 10^{-19} \text{ J*nm})(Q_1 Q_2 / r)$$

where E is in joules, r is in nanometers, and Q_1 and Q_2 are ion charges. Ion pairs have lower energy than ions that are separated. If the energy is negative, the attraction is positive. If the energy is positive, the attraction is repulsive.

Let us take a look at the bonding force between identical atoms such as hydrogen. When hydrogen atoms are brought together, there are proton-proton repulsion, electron-electron repulsion, and proton-electron attraction. Bond formation results when the system goes to the lowest energy state. Both atoms will position themselves to minimize repulsive and attractive terms. Bond length is where the energy is at a minimum. The net potential energy has to do with the attractions and repulsions of charged particles and kinetic energies. Atoms are infinitely far away from each other when the energy is zero. Repulsive forces occur when the atoms are very close together and their energy rises. The minimum energy is where bond length takes place.

For the H_2 molecule, electrons are in the space between the nuclei. This is where the electrons are attracted by the protons. Thus we find that the H_2 molecule is stable. According to the graph, we find the potential energy is smaller because of attractive forces. A bond formed between two hydrogen atoms makes the molecule stable. Proton-proton and electron-electron repulsive forces become balanced according to the hydrogen molecule's bond length. Electrons that are shared by nuclei are called covalent bonding. Ionic bonding involve oppositely charged ions being attractive to each other. Covalent bonding involves identical atoms sharing electrons. Polar covalent bonding involve electrons being transformed, but there seems to be unequal sharing of electrons. There is a bond polarity in such polar covalent bonding molecules

Electronegativity

Electronegativity is the atom's ability to attract electrons that are shared. Let us take a look at a molecule by the general formula HX. We can use the formula

Bond energy for H-X = [Bond Energy (H-H) + Bond Energy (X-X)]/2

The molecule H-X is thought of as having ionic and covalent bonding. The attraction between these atoms will give a greater bond strength. The more of a distance in electronegativities of atoms, the greater the ionic component and the greater the bond energy value. Relative electronegativities of H and X is obtained from bond energy values. Electronegativity values have been obtained for the elements using this process. Electronegativity normally increases from bottom to top and normally increases from left to right. Identical atoms share electrons equally resulting in a zero electronegativity value. This gives a covalent bond. An intermediate value of electronegativity gives a polar covalent bond. A large difference in electronegativity gives rise to an ionic bond.

Bond Polarity

Molecules that have a positive and negative center of charge are said to be dipolar. In other words, they have a dipole moment. An arrow points to the negative charge center while the tail points to the positive center. Dipole moments are found in both covalent and polar covalent bonds for diatomic and even polyatomic molecules. Water and ammonia each have unique dipole moments. The sum of each of the dipoles for individual bonds gives a net dipole moment for each of the molecules. It is possible for molecules to have polar bonds and not have a dipole moment. This happens when the bond polarities are arranged so that they cancel each other out. The molecule carbon dioxide is an example of this. The bond polarities cancel each other out because the molecule assumes a linear shape of equal size. Other examples of polar bonds with no dipole moments include sulfur trioxide and carbon tetrachloride.

Size

Electron configurations describe what makes a stable compound. The atoms have a noble gas arrangement of electrons. It is possible for nonmetallic elements to obtain noble gas electron configurations. They can share electrons with other nonmetals to make covalent bonds. It is also possible for nonmetals to take electrons from metals in order to form ions. Nonmetals have a tendency to form anions while metals have a tendency to form cations. Nonmetals react to make a covalent bond. They attain noble gas configurations. Nonmetals and metals react with each to form binary ionic compounds. Nonmetals achieve electron configurations of a noble gas. The valence orbitals of the metal remain unfilled. The nonmetal that forms the anion and the metal that forms the cation achieve noble gas electron configurations.

Ionic compounds are usually stated as the solid state. Positive and negative ions are held together by positive and negative attractions. In the gas phase, the ions are far apart from each other. Formulas of ionic compounds are electrically neutral. The positive and negative charges balance out to be zero. Metals lose 1 electron in Group 1A, 2 electrons in Group 2A, and 3 electrons in Group 3A. Nonmetals gain 1 electron in Group 7A and they gain 2 electrons in Group 6A. Hydrogen can either lose or gain an electron to form either the hydronium ion (H^+) or the hydride ion (H^-). There are other elements whose chemistries are different. Tin forms the ions Sn^{2+} and Sn^{4+}. Lead forms the ions Pb^{2+}

and Pb^{4+}. Bismuth forms the ions Bi^{3+} and Bi^{5+} and thallium forms the ions Tl^+ and Tl^{3+}. Transition metals form a variety of ions.

The size of an ion helps us understand structure, stability, properties, and its biological effects. It is not easy to define the size of ions. Ionic radii are determined between ion centers. Positive ions are formed by removing one or more electrons from an atom that is neutral. The cation itself becomes smaller than the parent atom. Adding electrons to an atom that is neutral gives an anion that is larger than the parent atom. The size of the ions and their position on the periodic table is important. The size of an ion increases as we go down a group. As we go horizontally across a row, we find there is no general trend. Metals give up electrons to form cations which become smaller and nonmetals accept electrons to form anions which become larger. Let us also look at the sizes of isoelectronic ions. These are the ions that contain the same number of electrons. The relative size of the ions have to do with the number of electrons and the number of protons. The size of anions are larger while the size of cations are smaller. The electrons in cations experience stronger attraction on the positive charge of the nucleus. This is why cations are smaller. For isoelectronic ions, size decreases as nuclear charge increases.

Binary Ionic Compounds

Metals and nonmetals react with each other to form cations and anions that are attractive. Ionic solids form because the oppositely charged ions have a lower energy than the separate elements themselves. The strength of the ions attracted to each other has to do with the lattice energy. We define the lattice energy as that energy change when separate gas ions come close together to form an ionic solid. Lattice energy is that energy that is released when the ionic solid forms. Lattice energy is considered to have a negative sign. Let us consider energy changes dealing with ionic solids forming by considering a hypothetical reaction.

$$M(s) + (1/2)X_2(g) \longrightarrow MX(s)$$

Note that energy is considered to be a state function. We are going to look at the steps and sum all of the steps to give an overall reaction.

Step 1: Sublimation
$M(s) \longrightarrow M(g)$

Step 2: Ionization
$M(g) \longrightarrow M^+(g) + e^-$

Step 3: Dissociation
$(1/2)X_2(g) \longrightarrow X(g)$

Step 4: Formation of Anions

$X(g) + e^- \longrightarrow X^-(g)$

Step 5: Formation of an Ionic Solid

$M^+(g) + X^-(g) \longrightarrow MX(s)$

Sum of the Reactions

$M(s) + (1/2)X_2(g) \longrightarrow MX(s)$

The ions are thought of as hard spheres packing together by maximizing attractions of oppositely charged ions and minimizing repulsions of identical ions that are charged.

Lattice Energy

Lattice energy contributes to stability of ionic solids. Lattice energy is represented by the equation

$$\text{Lattice Energy} = k(Q_1 Q_2/r)$$

where k is a proportionality constant, Q_1 and Q_2 are charges of the ions, and r is the distance between the centers of cations and anions. Lattice energy has a negative sign since Q_1 and Q_2 have opposite charges. Cations and anions are bought together due to the exothermic process. Energies involved in forming solid ionic compounds tell us that there are factors involved in determining the composition and structure of compounds. Balancing of these energies to make highly charged ions and energy being released when these ions form a solid are important.

Characteristics of Bonds

Atoms with different electronegativities and electrons not being shared equally gives a polar covalent bond. However, if there is a large difference in electronegativity, complete transfer occurs to form individual ions. Let us take a look at the percent ionic character for bonds of binary compounds in the gas phase. We can calculate the percent ionic character of a bond.

% Ionic Character =
(Measured Dipole Moment of X-Y/Calculated Dipole Moment of X^+Y^-) x 100%

Ionic character increases as the difference in electronegativity values increase. Normally, the bonds never reach 100% ionic character. Individual bonds cannot be completely ionic. Results of values such as these are for the gas phase where single bond molecules exist. The solid state is more complicated since multiple ion interactions are involved. Compounds that have an ionic character of 50% or greater is considered to be ionic. Ionic compounds also contain polyatomic ions. Keep in mind that these compounds are still ionic, but the individual ions are made up of covalent bonds.

We can define ionic compounds as compounds that conduct electricity when they are melted and dissolved in water.

Chemical bonds can be thought of as groups of atoms held together as a unit. Bonds form because the system wants to be in the lowest possible energy state. They form when the atoms are more stable and have lower energy when the atoms are separate. We can think of a chemical bond as a model based on molecular stability. Methane has the general formula CH_4. The hydrogens are at the corners of a tetrahedron around the carbon atoms. Bonds written for molecules is a concept chemists made up. It becomes a method for calculating the energy produced when a stable molecule is formed from individual atoms. We can think of bonds as a quantity of energy from an energy of stabilization.

We use models in chemistry to explain what is happening in the microscopic world based on the macroscopic world. We further our understanding of chemistry by studying those models and how they are used to help us. Bonding is thought of as a model to understand how stable molecules are. Bonds come from the idea that chemical processes involve atoms being held together and reactions are ways atoms are arranged. Processes tend to go through lower energy states. Atoms held together are in a lower energy state than atoms that are separated from each other. The reason is atoms share electrons or atoms transfer electrons and become ions. Bonds form between atoms with a certain value of energy per mole. Bonding models help us explain chemical behavior by considering molecules as collections of components. We tend to think of a bond as behaving the same in any type of molecular environment. Atoms tend to form stable groups and they do this by sharing electrons. A lower energy state is achieved.

Bond Energies

Let us look at the energy of different types of bonds. Let us also see how the bonding concept deals with energies of chemical reactions. The sensitivity of a bond to its environment can be understood from experimental measurements of energy required to break a C-H bond. The C-H bond strength varies based on its environment. One pair of electrons being shared is considered to be a single bond. Atoms that share two pairs of electrons is called a double bond, while atoms with three pairs of electrons form triple bonds. Note that as electrons being shared increases, bond length gets smaller. Values of bond energy is used to calculate energies for reactions. Energy is added to the system in order to break bonds. This is considered to be an endothermic process. The sign is positive. Energy is released from the system by forming a bond. This is considered to be an exothermic process. The sign is negative. The equation for the enthalpy change for a reaction is

$$\Delta H = \sum D(\text{Bonds Broken}) - \sum D(\text{Bonds Formed})$$

$$\Delta H = \text{Energy required} - \text{Energy released}$$

The symbol E gives us the sum of terms and D gives us the bond energy per mole of bonds. Average bond energy values are obtained from tables.

An Electron Bonding Model

The chemical bonding model involves properties like bond strength and polarity to individual bonds. But we want to describe covalent bonds. The localized electron model bases its assumptions that a molecule is made up of atoms that share pairs of electrons using atomic orbitals. Electron pairs are on localized atoms, or they can be in the same space between atoms. Lone pairs are electrons on the atom. Lone pairs are also called nonbonding pairs or non sharing pairs. Those formed between the atoms are called bonding pairs. The localized electron model involves describing the valence electron arrangement using Lewis Structures, help describe the geometry using the valence electron-pair repulsion model, and it describes what kind of orbitals are used.

Lewis Structures

The Lewis Structure tells us how the valence electrons are arranged among atoms in a molecule. Stable compounds involve atoms with noble gas electron configurations. Metals reacting with nonmetals form binary ionic compounds by transferring an electron from the metal to the nonmetal. Noble gas electron configurations have been achieved. The metal ion becomes a cation and a nonmetal becomes an anion. The cation achieves the previous noble gas electron configuration whereas the anion achieves the next noble gas electron configuration. The duet rule states that two electrons are shared. An example of this is the hydrogen molecule. There are two electrons between the hydrogen atoms that form a stable molecule. The element helium has a filled noble gas electron configuration. Its valence orbital has already been filled. Nonmetals in the second row can form stable molecules if they are surrounded by enough electrons for the valence orbitals to be filled. Eight electrons are needed to fill in the orbitals. They obey the octet rule. The elements carbon, nitrogen, oxygen, and fluorine tend to obey the octet rule when stable molecules are formed. The element neon obeys the octet rule by having eight valence electrons around it. It does not form bonds. The steps to writing Lewis Structures is to add up the number of valence electrons from all atoms, then using a pair of electrons to form a bond between the atoms, and making sure the duet rule is used for hydrogen and the octet rule is used for second-row elements. Lines are used to show each pair of bonding electrons. Molecules are thought of as entities that use valence electrons to achieve a lowest possible energy. The valence electrons are part of the whole molecule.

Keep in mind there are exceptions to the octet rule. Boron forms compounds that have fewer than eight electrons. The boron atom is thought to be electron-deficient. An example of an electron deficient compound is BCl_3. The boron atom has an empty p orbital that is ready to accept electrons from a molecule. Either of the molecules NH_3 or H_2O can form a stable bond with the boron atom to achieve an octet of electrons. The elements carbon, nitrogen, oxygen, and fluorine normally obey the octet rule. There are also atoms that exceed the octet rule. These elements are normally in period 3 and beyond. An example would be SCl_6. The element chlorine obeys the octet rule, but sulfur has 12 electrons around it. If we take a look at the valence orbitals for the second and third period elements, we see there are some differences. Second-row elements have 2s and 2p valence orbitals. Third-row elements have 3s, 3p, and 3d orbitals. There are empty 3d orbitals that can be used to fill in extra electrons. Eight of the electrons are used to fill in the 3s and 3p orbitals. The extra four electrons are filled in the empty 3d orbitals. Extra electrons are placed on the central atom for third-row or elements beyond. In this case, it is okay to exceed the octet rule. An example would be I_3^-. Molecules such as nitric oxide and nitrogen dioxide are considered to be odd electron molecules. These are also exceptions to the octet rule.

Resonance

More than one kind of Lewis Structure can be drawn for a molecule. The correct way to describe resonance structures is a blending or superposition of all three or more of its structures. Polyatomic ions are normally written for resonance structures. We can think of the structures as the average of multiple structures drawn. Resonance is a term that is used to describe more than one Lewis Structure being written for a molecule. An electron structure of the molecule itself is a resonance structure. Double-headed arrows are used to represent one molecule from another. The nuclei or atoms are fixed in space and are arranged the same. The only thing that is different is how these electrons are placed. The arrows used to represent resonance structures can seem to be misleading. They do not mean one molecule flips back and forth from one molecule to another. The actual structure is an average of all the resonance structures. Electrons are not really localized like the Localized Electron Model. Electrons are actually delocalized. They move around the entire molecule. We can think of resonance as a better description than the Localized Electron Model.

Formal Charge

Molecules that exceed the octet rule have different kinds of nonequivalent Lewis Structures. But they still obey the rules for writing Lewis Structures. The Lewis Structures that describe the bonding in molecules or polyatomic ions is best estimated based on charge. The charge on each of the atoms determine the appropriate or best possible structure. Atomic charges need to be assigned on the molecules when they are written. The formal charge is the number of valence electrons minus the number of valence electrons given to an atom in a molecule. Determining the formal charge of an atom requires us to know the number of valence electrons on an atom that is neutral, and the number of valence electrons that belong to an atom. The numbers are then compared. If the number of valence electrons is the same as in the free state, the positive and negative charges balance each other. The formal charge is zero. If there is more than one valence electron, the formal charge is -1.

When computing formal charge, lone pair electrons belong to the atom and shared electrons are equally divided. Formal charge can be defined as the following

$$\text{Formal Charge} = (\text{\# of Valence e}^-) - (\text{\# of Lone Pair e}^-) - (1/2)(\text{\# of Shared e}^-)$$

When evaluating Lewis Structures, atoms achieve formal charges near zero or as close as possible. Formal charges that are negative are on electronegative atoms, and formal charges that are positive are on electrophilic atoms. Note that formal charge gives an estimate of change.

The VSEPR Model

The VSEPR Model has to do with molecular structure. Structures of molecules help us determine their chemical properties. Molecular structure can be thought of as three-dimensional arrangement of atoms. We can also predict the approximate molecular structure by using the Valence Shell Electron-Pair Repulsion (VSEPR) Model. This model helps us predict the geometry of molecules that are formed from nonmetals. The structure around an atom is determined by minimizing electron-pair repulsions. Bonding and nonbonding pairs should be positioned as far apart as possible. The molecule carbon dioxide has a linear structure with bond angles of 180 degrees. The molecule boron trichloride has a trigonal planar structure with bond angles of 120 degrees. The molecule methane has a tetrahedral structure with bond angles of 109.5 degrees. The goal of the VSEPR Model is to have the electron pairs arranged around a central atom that minimizes the repulsions. The molecule's structure can be determined. The name of the structure is based on the position of the atoms and where the electron pairs are arranged. Ammonia is an example of a tetrahedral arrangement of electron pairs. However, the correct molecular structure is a trigonal pyramid. The bond angle in methane is 109.5 degrees. The bond angle for ammonia is 107 degrees. The bond angle for water is 104.5 degrees. Bond angles decrease as the number of lone pairs increase. Lone pairs take up so much space compared to bonding pairs. More lone pairs cause bonding pairs to be squeezed together.

Bonding pairs are shared between two nuclei. Electrons are close to the nucleus. Lone pairs are localized on one nucleus, and they will be close to one nucleus. Lone pairs take up a lot of space than bonding pairs, and they compress angles between them. The molecule phosphorus pentabromide has a trigonal bipyramid structure. The structure is made up of two trigonal-based pyramids sharing a common base. The anion phosphorus hexabromide has an octahedral structure. The molecule xenon tetrafluoride has a square planar structure. The nitrate anion has a trigonal planar structure. According to VSEPR theory, the double bond is counted as one pair. The electron pairs are in the space between nuclei to form a double bond. The double bond acts as a center of electron density that repels nearby electrons. Multiple bonds are treated as one electron pair according to the VSEPR Model. Any resonance structure can be used to predict the molecular structure based on the VSEPR Model. The water molecule has a V-shaped structure. The two lone pairs on oxygen push the bonding pair of electrons to form an angle less than 109.5 degrees. In fact, the bond angle is 104.5 degrees. Any size of the molecule's structure can be determined using the VSEPR Model to each atom. However, keep in mind there are always exceptions to any simplified model such as VSEPR.

Conclusion

Chemical bonds hold atoms together. Bonding happens when atoms lower the energy by coming together. Ionic bonds transfer electrons to form ions. Covalent bonds involve electrons being shared. Polar covalent bonds has to do with electrons being shared unequally. Percent ionic character for a diatomic molecule is the ratio of the measured dipole moment to its calculated dipole moment multiplied by 100%. Electronegativity is the ability of an atom to attract electrons that are shared in a bond. Electronegativity difference determines bond polarity. The arrangement of the polar bonds determine the molecule's overall polarity or dipole moment of a molecule. Stable molecules have their valence orbitals filled. Nonmetals bond and achieve noble gas configurations by covalent bonding. Noble gas configurations are also achieved by a metal transferring an electron or more to a nonmetal. Cations are smaller than its parent atom because it has lost one or more electrons. Anions that have more than one electron than a parent atom are considered to be larger. Normally, the size of an ion increases when we go down a group. Isoelectronic ions get smaller as the value of the atomic number increases. Lattice energy is a change in energy when gaseous ions come together to form an ionic solid. Bond energy is energy needed to break a covalent bond. The amount of energy needed has to do with the number of electron pairs that are shared. Single bonds have one electron pair. Double bonds have two electron pairs, and triple bonds have three electron pairs. Bond energies are also used to calculate enthalpy change for a reaction. Lewis Structures tell us how the valence electrons should be arranged among atoms. Atoms fill their valence orbitals. Hydrogen obeys the duet rule. Second-row elements obey the octet rule. Third-row elements and beyond can fill in more electrons than eight because they have d orbitals. It is possible for more than one Lewis Structure to be drawn. This is called resonance and the actual structure of a molecule is a combination or blending of all resonance structures. The molecule with the lowest formal charge is the best Lewis Structure written based on many nonequivalent structures. VSEPR stands for Valence Shell Electron Pair Repulsion, and it is used to predict geometries of a molecule by nonmetals forming. The best structure possible is by minimizing electron-pair repulsions.

Chapter 10
Molecular Orbitals

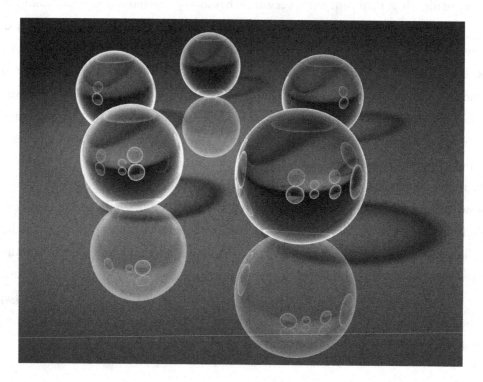

Bubbles

Hybridization

The localized electron model is thought of as bunch of atoms that are held together by sharing electrons between atomic orbitals. Valence electrons are used to show the Lewis Structure, and its molecular geometry is based on the VSEPR model. Atomic orbitals will be described and used to show how shared electrons form bonds.

sp³ Hybridization

Bonding involves the valence orbitals. Let us take a look at methane. Hydrogen atoms use 1s orbitals. Carbon atoms use 2s and 2p orbitals. By experiment, methane is known to be tetrahedral that has bond angles of 109.5 degrees. The carbon atom has four equivalent atomic orbitals in a tetrahedral arrangement. Mixing of atomic orbitals is considered to be hybridization. The four new orbitals that are formed is considered to be sp^3. There is one 2s and three 2p orbitals. Carbon is considered to be sp^3 hybridized. The sp^3 orbitals have the same shape. Orbital energy-level diagrams are used to describe hybridization. We are not concerned about electron arrangements on the atoms. The total number of electrons and its arrangement is important. The atomic orbitals for carbon are arranged for the best electron arrangement for the molecule. The new atomic orbitals on carbon are used to share electron pairs with 1s orbitals from four hydrogen atoms. Atomic orbitals that form a tetrahedral structure is considered to be sp^3 hybridized. Atoms adopt hybrid orbitals to achieve a minimum energy state. Molecules as a whole are more than their individual parts.

sp² Hybridization

Ethylene is a simple molecule that is used to make plastics. The molecule has a total of 12 valence electrons, and it contains a trigonal planar arrangement that has bond angles of 120 degrees. Hybrid orbitals are used to form bonds. Carbon uses the 2s and 2p valence electrons whereas hydrogen uses 1s valence electron. Three orbitals that are arranged at 120 degree angles combines an s orbital and two p orbitals. It takes one 2s and two 2p orbitals to form hybrid orbitals. This is called sp^2 hybridization. One of the 2p orbitals on carbon has not been used. The p orbital in the z direction is perpendicular to the plane of the sp^2 orbitals. These orbitals are used to make bonds in ethylene. Three sp^2 orbitals on the carbon are used to share electrons. The chemical bonds have electron pairs between the atoms. We consider a covalent bond to be a sigma bond. The ethylene molecule we have been describing have sigma bonds involving sp^2 orbitals on each carbon atom and a 1s orbital on each hydrogen atom. An electron pair is also between carbon atoms. The other electron pair is found in the space above and below the sigma bonds. In this case, the 2p orbital is perpendicular to sp^2 hybrid orbitals. The p orbitals are parallel and shares an electron pair. The space is above and below the atom to form what is called a pi bond. Remember that a double bond contains a sigma bond and a pi bond. Carbon atoms contain sp^2 hybrid orbitals and form sigma bonds to hydrogen atoms and to each other. The p orbitals are used to form pi bonds to each other. The Lewis Structure for ethylene contains a carbon-carbon double bond and carbon-hydrogen single bonds.

sp Hybridization

The molecule carbon dioxide has sp hybridization with a bond angle of 180 degrees. The atomic orbitals are in opposite directions. Hybridization of sp involves an s orbital and a p orbital. Let us take a close look at the bonding in carbon dioxide. There are sp orbitals on carbon that form sigma bonds and there are sp^2 orbitals on the oxygen atoms that have lone pairs. There are pi bonds between carbon and oxygen that are formed by the overlap of 2p orbitals. The carbon atom that is sp hybridized contains two unhybridized p orbitals. The p orbitals form a pi bond with oxygen.

dsp^3 Hybridization

The molecule phosphorus pentabromide has dsp^3 hybridization. The molecule has a trigonal bipyramidal arrangement. Five of the P-Cl sigma bonds form due to sharing of the electrons. The dsp^3 orbital on the phosphorus atom and a sp^3 orbital bond to each other.

d^2sp^3 Hybridization

The molecule sulfur hexachloride is an example of a molecule that has a d^2sp^3 hybridization. The chlorine atoms form an octahedral arrangement around the sulfur atom. An octahedral set of six orbitals are also formed. There are two d orbitals, one s orbital, and three p orbitals for d^2sp^3 hybridization. The d^2sp^3 orbitals of sulfur bonds to each of the chlorine atoms. The chlorine atoms are sp^3 hybridized.

Table 10-1: Information About the Shapes of Molecules

Number of Bonds	Geometry	Hybridization	Angle	Molecule
2	Linear	sp	180°	$BeCl_2$
3	Trigonal Planar	sp^2	120°	BCl_3
4	Tetrahedral	sp^3	109.5°	CH_4
5	Trigonal Bipyramidal	dsp^3	120°, 90°	PCl_5
6	Octahedral	d^2sp^3	90°	SCl_6

The Molecular Orbital Model

The localized electron model is important when studying the structure and bonding of molecules. However, the model makes incorrect assumptions. In other words, it assumes that electrons are localized. Resonance needs to be added to the picture. This model does not deal with molecules that have unpaired electrons. The model also does not give information that deal with bond energies. The molecular orbital model deals with bonding. Let us take a look at the H_2 molecule. The hydrogen molecule contains protons and electrons. Note the atomic theory assumes electrons in the atom are in orbitals of specific energy. The electron correlation problem does not tell us the details of electron movements. It is, therefore, not easy to understand electron-electron interactions. Approximations of solutions of a problem should be made without ruining the concept of the model itself. These approximations are measured based on theory with experimental observations. Atomic orbitals are solutions of atoms. Molecular orbitals are also solutions to molecular problems. Both types of orbitals hold two electrons with spins that are opposite from each other. The square of the wave function gives electron probability.

Let us describe the bonding in the hydrogen molecule. Molecular orbitals are made from the hydrogen 1s atomic orbitals. When equations for the hydrogen molecule is solved, molecular orbitals are formed. They can be represented as:

$$MO_1 = 1s_A + 1s_B$$

$$MO_2 = 1s_A + 1s_B$$

Size, shape, and energy are important when describing orbital properties. Let us take a look at what we know for hydrogen molecular orbitals. The electron probability of the molecular orbitals is between two nuclei. For the first molecular orbital, most of the electron density is between the nuclei. For the second molecular orbital, it can be on either side where the nuclei is at. Both types of orbitals are considered to be sigma molecular orbitals. Molecular orbitals are available for electrons. The first molecular orbital is in a lower energy state than orbitals of free hydrogen atoms, whereas the second molecular orbital is higher in energy than 1s orbitals. The electrons occupy lower energy molecular orbital than two separate hydrogen atoms. Molecule formation is then formed. Nature tends to direct itself towards a lower energy state. This is important for bonding.

Let us suppose the two electrons occupy the higher molecular orbital. In this case, this would be the second molecular orbital. This would be considered to be anti bonding. Electrons should be lower in energy than separated atoms. The lower state is favored over the anti bonding state. The two electrons are better occupied in the lower energy first molecular orbital because of stability. The hydrogen molecule has two types of orbitals. The bonding molecular orbital is in a lower energy state than the atomic orbitals. Electrons prefer to be in this orbital and bonding is favored. An anti bonding molecular orbital is in a higher energy state than the atomic orbitals. Electrons like to be as separated atoms. The bonding molecular orbital for H_2 has their electrons a better probability between nuclei. Electrons lower their energies by being attracted to two nuclei. Electrons for the anti bonding molecular orbital are normally outside the nuclei. This does not provide any bonding force. Electrons are in a higher energy state of the atom. Molecular orbitals are labeled to tell us their shape or symmetry, the atomic orbitals, and if they are bonding or anti bonding. An asterisk is used to show anti bonding. The H_2 molecule has sigma symmetry. They are made from hydrogen 1s atomic orbitals.

$$MO_1 = Sigma(1s)$$

$$MO_2 = Sigma*(2s)$$

Molecular electron configurations are written like atomic electron configurations. The H_2 molecule has two electrons and these electrons are in the Sigma(1s) molecular orbital. The electron configuration can be written as $Sigma^2 1s$. Molecular orbitals have two electrons, and they have spins that are opposite from each other. The number of molecular orbitals formed should equal the number of atomic orbitals to make them.

Bond Order

Bond order tells us about bond strength. Bond order can be defined as the number of bonding electrons and the number of anti bonding electrons being subtracted from each other and then divided by two.

Bond Order = (Number of Bonding Electrons - Number of Antibonding Electrons)/2

The number 2 is divided because bonds are based on pairs of electrons. A larger bond order tells us the strength of the bond is greater. A smaller bond order tells us the strength of the bond is less.

Bonding in Homonuclear Diatomic Molecules

Homonuclear diatomic molecules are made up of two identical atoms. They are from the elements in Period 2 of the periodic table. Atomic orbitals are to be overlapped for there to be molecular orbitals. Valence orbitals contribute to molecular orbitals. Let us take a look at p orbitals. The two p orbitals combine with each other to make two sigma molecular orbitals. One of them is bonding and the other one is antibonding. It is also possible for two p orbitals to lie parallel from each other to make two pi molecular orbitals. One of them is bonding and the other one is antibonding. Electrons in bonding molecular orbitals are between nuclei. Electrons in anti bonding molecular orbitals are outside the area between the two nuclei. These are considered to be sigma molecular orbitals. When we are looking for p orbitals that lie parallel with each other, they produce bonding and anti bonding orbitals. The electron probability is above and below between the nuclei. These are considered to be pi molecular orbitals. They are labeled as pi_{2p} for bonding molecular orbital and pi^*_{2p} for anti bonding molecular orbital. Let us take a look at the energies of sigma and pi molecular orbitals that are formed from 2p atomic orbitals. The sigma orbital is lower in energy since electrons are closest to two nuclei. The interactions of sigma bonds are stronger than the interactions of pi bonds.

Paramagnetism

When there is a field present, there are two types of magnetism that can be induced. Paramagnetism makes a substance attracted to the inducing magnetic field. Diamagnetism makes a substance repelled from an inducing magnetic field. Paramagnetism has to do with unpaired electrons, and diamagnetism has to do with paired electrons. A substance containing both paired and unpaired electrons shows a net paramagnetism. The effects of paramagnetism is a lot stronger than diamagnetism. The diatomic molecule boron is paramagnetic with a bond order of 1. The diatomic molecule carbon is diamagnetic with a bond order of 2. The diatomic molecule nitrogen is diamagnetic with a bond order of 3. The diatomic molecule oxygen is paramagnetic with a bond order of 2. The diatomic molecule fluorine is diamagnetic with a bond order of 1. Information such as the following gives us an idea about bond energy, bond length, and bond strength. As bond order increases, bond energy increases and bond length decreases while bond strength increases. As bond order decreases, bond energy decreases and bond length increases while bond strength decreases.

Bonding in Heteronuclear Diatomic Molecules

When two atoms of a diatomic molecule are different, a different energy-level diagram is used. An example of a heteronuclear diatomic molecule would be NO. The molecular orbital electron configuration would then be determined. The bond order will come out to be 2.5. In fact, we find that nitric oxide is paramagnetic experimentally in agreement with the unpaired electron according to the molecular orbital electron configuration.

Combination of both Models

The localized electron model assumes electrons are localized. However, this does not work since some molecules require more than one Lewis Structure to be drawn. Not one single structure describes the overall structure of the molecule. The blending or hybrid mixing of the structures is a more accurate way to describe the structure of a molecule. The term resonance is used in this fashion. But even resonance included with the localized electron model is insufficient to describe a molecule. If we combine the localized electron model with the molecular orbital model, we can describe the bonding differently. If we examine the resonance structures for O_3 and NO_3^- we find that the sigma bonds can be described in terms of the localized electron model. The pi bonding is treated as being delocalized. When we are dealing with resonance, the localized electron model describes sigma bonding and the molecular orbital model describes pi bonding.

Let us take a look at benzene. Benzene has the formula C_6H_6 and it consists of a planar hexagon. There are six carbon atoms arranged in a hexagonal-like structure where each carbon atom bonds to one hydrogen atom. All carbon to carbon bonds are known to be equivalent. There are two resonance structures for benzene. We can also describe the bonding in benzene if we combine models. Sigma bonds of carbon atoms involve sp² orbitals. Sigma bonds are in the plane of the molecule. There are six p orbitals that are combined to form 6 sp² hybridized carbon atoms. Electrons in the formation of pi molecular orbitals are delocalized over the ring of carbon atoms. This gives six equivalent bonds. The benzene structure can also be written as a hexagon with a circle inside it. This is used to show delocalized bonding in the molecule.

Conclusion

Covalent bonding has an important role for orbitals. The localized electron model and the molecular orbital model are both used to describe the structure of molecules. The localized electron model describes group of atoms that share electron pairs using atomic orbitals. New hybrid orbitals are formed. The orientation of these hybrid orbitals depends on the electron pairs giving a minimum repulsion. Tetrahedral arrangements have sp³ hybrid orbitals, trigonal planar arrangements have sp² hybrid orbitals, and linear arrangement have sp hybridization. A sigma bond is a covalent bond with an electron pair. Pi bonds occur when p orbitals overlap in a parallel way. Electron pairs occupy a region of space above and below the bonded atoms. A double bond has one sigma and one pi bond. A triple bond has one sigma and two pi bonds. Multiple bonds have both sigma and pi bonds.

The molecular orbital for covalent bonding consists of positively charged nuclei with electrons. Molecular orbitals are made from valence orbitals of atoms forming a molecule. Molecular orbitals are classified based on their energy and shape. Bonding molecular orbitals are lower in energy than atomic orbitals, whereas anti bonding molecular orbitals are higher in energy than the same atomic orbitals. Sigma molecular orbitals have electron probability between two nuclei. Pi molecular orbitals have electron probabilities above and below the nuclei. Bond order gives us an idea of bond strength. The higher the bond order, the greater the stability of the molecule. The advantages of the molecular orbital model tells us the relative bond strength and the kind of magnetism in diatomic molecules, bond polarity, and it gives us a description of electrons being delocalized in polyatomic molecules. The disadvantage of the molecular orbital model is being able to apply it to polyatomic molecules. Resonance is introduced to molecules to show that one or more Lewis Structures can be drawn. Combining the localized and molecular orbital models gives us a better description of molecules than either one alone. Sigma bonds are thought as being localized, whereas pi bonding is thought of as being delocalized.

Chapter 11

Intermolecular Forces

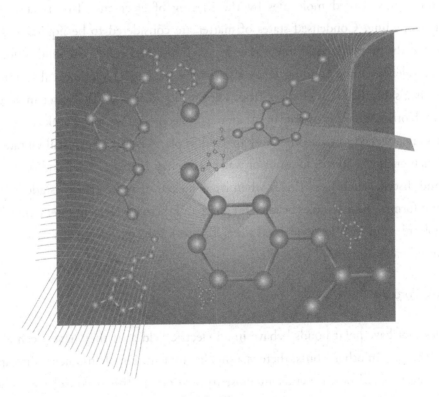

Molecule Illustration

Water is arranged differently in its gas, liquid, and solid forms. Molecules of gaseous water molecules are far apart from each other. They have low density, high compressibility, and it also completely fills a container. Water molecules in the liquid state stay close to each other, but they are not far apart nor are they rigid. Water in its solid state is ice. Ice is less dense than water because of empty spaces inside its structure. It is also less dense than liquid water because of special types of forces that will be discussed later on. Normally, solids have much greater densities than liquids. They are very compressible, and they are rigid. Solids keep their shape regardless of the container. Molecules in a solid are close to each other and have large attractive forces. If we take a look at enthalpy values for changes of state for water we find that the following values hold experimentally

$$\Delta H_f = 6.02 \text{ kJ/mol}$$

$$\Delta H_v = 40.7 \text{ kJ/mol}$$

The ΔH_f is the enthalpy of fusion, and ΔH_v is the enthalpy of vaporization. There seems to be a bigger change going from a liquid to a gaseous state than a change going from a solid to a liquid state. This tells us that there are a lot of attractive forces in liquid water. Densities of the solid state and liquid state of water are close to each other. Let us take a look at the properties and structures of a liquid and a solid. We will also take a look at changes of state that happen between solid and liquid, liquid and gas, and solid and gas.

Atoms form units called molecules by the sharing of electrons. This is considered to be intramolecular bonding. Condensed states of matter are considered to be liquids and solids. Let us take a look at the properties and forces that form a liquid or a solid. The kind of forces that are involved are covalent or ionic bonding. They also involve weaker interactions called intermolecular forces. Water is a substance that changes from solid to liquid to gas. The water molecules remain attracted to each other. Changes of state occur among molecules. If we take a look at ice, for example, the molecules are attached to each other and are held in place. But they can still vibrate about their positions. When energy is added, motions of water molecules increase. They end up having greater movement and disorder increases. This is when ice is melted. If more energy is added, the gaseous state exists and forms. The molecules become far apart and there is negligent interaction. The gas still has water molecules but is in its condensed state. To break the covalent bonds in a water molecule requires a considerable amount of energy.

Dipole-Dipole Forces

Molecules that have polar bonds behave in an electric field as if they have a center of positive and negative charges. In other words, there is a dipole moment. Dipole moments line up with each other so the negative and positive ends are close to each other. This is considered to be a dipole-dipole attraction. In the liquid state, molecules do their best to be with each other by attraction or repulsion. The molecules arrange themselves so the positive and negative attractions are together and to minimize the positive-positive and negative-negative interactions.

Dipole-dipole forces become weaker as the dipoles further increase from each other. Molecules are far apart from each other and their forces are not important in the gas phase. There are strong dipole-dipole forces when hydrogen is attached to a highly electronegative atom. In this case, the atom could be nitrogen, oxygen, or fluorine. The strengths of these interactions depend on the polarity of the bond and how close the dipoles are to each other that is allowed by the small size of a hydrogen atom. These dipole-dipole attractions are considered to be strong. They are given the name hydrogen bonding. Hydrogen bonding among water molecules happen between partially positive hydrogen atoms and lone pairs on adjacent water molecules.

Hydrogen bonding has an effect on physical properties. Large interactions of hydrogen bonding exist among small molecules with most polar X-H bonds. Strong hydrogen bonding forces are due to two reasons. One reason has to do with the large electronegativity of light elements in each group. This gives polar X-H bonds. Another reason has to do with the small size of the first element for each group. This allows dipoles to approach each other making the intermolecular forces stronger. Molecules that hydrogen bond to each other normally have high boiling points.

London Dispersion Forces

There are molecules without dipole moments that exert forces on each other. All substances including the noble gases exist in the liquid and solid state. Noble gas atoms and nonpolar molecules have London Dispersion Forces. Let us take a look at noble gas atoms. The electrons are assumed to be uniformly distributed around the nucleus. It is possible for there to be an electron distribution that is not symmetrical that makes a temporary dipolar arrangement of charge. These temporary dipoles affect the electron distribution of an atom that is nearby. An instantaneous dipole in an atom induces a similar dipole in an atom that is nearby. There is an interatomic attraction that is weak and does not stay for a very long time, but that plays an important role for large atoms. The interactions have to be strong enough to make a solid and the motions of the atoms must be slow. This is the reason why noble gas elements have low freezing points.

For Group 8A elements, the freezing point increases down a group. In this case, atomic number increases while the number of electrons increase. There is an increase of momentary dipole interactions. This can be described by polarizability. This indicates the case of an electron "cloud" of an atom is distorted to give a dipolar charge distribution. Large atoms that have a lot of electrons have a higher polarizability than small atoms. As the size of the atom increases, London Dispersion forces increase too. Dispersion forces with large atoms are more important than dipole-dipole forces. Keep in mind small nonpolar molecules like O_2, CH_4, CF_4, and CO_2 do not have dipole moments that are permanent. They attract each other through London Dispersion forces.

The Liquid State

Liquids are important for our everyday survival. We consider water to be the most important liquid. It provides a medium for food preparation, transportation, and a cooling mechanism for machines and industrial processes. It is also used for cleaning, recreation, and many other uses. Liquids have a lot of characteristics. They have low compressibility, not rigid, and they have high density when we compare them to other gases. The properties of a liquid give us an idea about the forces among the particles. If we were to pour a liquid on a surface, the water beads as droplets. This depends on intermolecular forces. Molecules in the interior of a liquid are surrounded by other molecules. Molecules at the surface of a liquid are attracted from the sides and from below. An uneven pull on the surface of the molecules draws them in the body of the liquid. This makes a droplet of liquid to be a sphere or a shape close to a sphere. This shape is considered to be the minimum surface area. In order to increase a liquid's surface area, molecules move from the interior of a liquid to its surface. Energy is required in order to overcome intermolecular forces. Surface tension is considered to be the resistance of a liquid that has been increased. Polar molecules have large intermolecular forces and have high surface tensions.

Polar liquids have the characteristic of capillary action. Capillary action is when a liquid rises in a narrow-tube. Two kinds of forces are responsible for capillary action. Cohesive forces are considered to be intermolecular forces among molecules of a liquid. Adhesive forces are forces between liquid molecules and the container itself. Adhesive forces happen when the container is composed of a substance that has polar bonds. An example would be a glass surface that has many oxygen atoms. The oxygen atoms have partial negative charges and are attracted to a positive end of a polar molecule like water. Water makes the glass wet by going up the walls of the tube. The surface of the water touches the glass. This increases the surface area of the water. This is opposed by cohesive forces that minimizes the surface. Water has strong intermolecular cohesive forces and strong adhesive forces to glass itself. This makes water go up in a glass capillary tube. It goes up to a specific height where the weight of the water balances the water attracted to the surface of a glass. If water was poured into a graduated cylinder, it forms a concave meniscus. A concave meniscus shows that water's adhesive forces are stronger than the cohesive forces. If mercury is poured into a graduated cylinder, a convex meniscus shape is formed. In this case, the cohesive forces are stronger than the adhesive forces.

Viscosity is a property of liquids that depend on intermolecular forces. It is a measure of a liquid to resist flow. Liquids that have large intermolecular forces are highly viscous. Glycerol is an example of a molecule that has high viscosity because of its capacity to form hydrogen bonds. It does this using hydroxyl groups. The more complex the molecule the higher its viscosity since large molecules become entangled. Gasoline is considered to be a nonviscous liquid that contain at least three to eight hydrocarbon molecules. Grease, on the other hand, is very viscous and has larger hydrocarbon molecules which can range anywhere from twenty to twenty-five. Spectroscopy is a technique that allows one to follow rapid changes that occur in liquids. Models of liquids are becoming more accurately defined. Liquids can be thought of as having large number of regions where the components are arranged like that in a solid, but there is more disorder. There is also a smaller number of places where the holes are present. It is a dynamic situation with rapid fluctuations in both regions.

Solids

Solids can be classified in two ways. They can be classified as crystalline solids or amorphous solids. Crystalline solids have their components highly arranged. Amorphous solids, on the other hand, have their components in greater disorder. Crystalline solids produce nice crystals that are beautiful. They do this because the arrangement of the components are regular. All positions of the components are represented by a lattice. A lattice can be thought of as a three-dimensional arrangement of points that show the positions of the components. A unit cell is considered to be the smallest repeating unit. A specific lattice can be made by repeating the unit cell to make a desired structure. The type of unit cells are simple cubic, body-centered cubic, and face-centered cubic. Also keep in mind there are noncrystalline materials that are amorphous. Glass, for example, is thought of as a solution where the components are mixed in place before making an arrangement that is in order. Glass has a rigid shape, but there is disorder in its structure.

The structure of a crystalline solid is determined by X-ray diffraction. A diffraction pattern is formed when there are beams of light being scattered from an array of points. The spacings between components are about the same order of magnitude as the wavelength of light. The parallel beams in phase results in constructive interference. The waves that are out of phase is due to destructive interference. X-rays of a single wavelength are targeted at a crystal where a diffraction pattern is formed. Light and dark areas of a photographic plate result because waves scattered from atoms can either reinforce or cancel each other. Waves can either reinforce or cancel each other due to the difference in distance traveled by waves after they strike atoms. Before the waves are reflected, they are in the same phase and the difference in distance gives an integral number of wavelengths. The waves will be in phase. The distance that is traveled depends on the distance between atoms. Therefore, the diffraction pattern is used to find out the interatomic spacings. Let us take a look at two different layers in a crystal where there are both incidental and reflected rays.

The lowest portion of the wave travels an extra distance and the sum of the two extra distances is comparable in magnitude. After reflection, the waves will be in phase given the sum of the two extra distances equals an integer times the wavelength of X-rays. The sum of the two extra distances is also equal to $2d\sin\Theta$ where d is the distance between atoms and Θ is considered to be the angle of incidence and reflection. The two equations combined together give $n\lambda = 2d\sin\Theta$. This is called the Bragg Equation. A diffractometer is an instrument used to carry out X-ray analysis of crystals. The crystal is rotated in relation to the X-ray beam and data is collected forming the scattering of X-rays from different planes of atoms in the crystal. The computer analyzes the results. X-ray diffraction can be used to gather data on bond lengths and bond angles and this can also predict models of molecular geometry.

Crystalline Solids

Different kinds of crystalline solids exist in nature. Ionic solids have their ions at points on their lattice. Molecular solids have molecules covalently bonded to each other. An example of an ionic solid would be sodium chloride. An example of a molecular solid would be sucrose. Another kind of solid is shown by elements such as carbon, boron, silicon, and metals. Atoms are at the lattice points, and they give a detailed structure of a solid. These kinds of solids are called atomic solids.

Buckminsterfullerene C_{60} is part of the fullerene family, and it is considered to be an atomic solid. Solids are classified based on the kind of components taking up and occupying the lattice points. Atomic solids have their atoms occupying the lattice points. Molecular solids have small molecules at the lattice points. Ionic solids have ions at the lattice points. Atomic solids can be subdivided into metallic solids, network solids, and Group 8A solids. Metallic solids have nondirectional covalent bonding that is delocalized. Network solids have their atoms bonded to each other with directional covalent bonds that make giant molecules of atoms. Group 8A solids have the noble gas elements attracted to each other by London Dispersion forces. Different bonding leads to different properties for the solids. Examples of atomic solids would be helium, gold, and graphite each having different properties.

Closest Packing

Metals are known to have high thermal and electrical conductivity, ductility, and malleability. They have these properties because of the nondirectional covalent bonding that is found in metallic crystals. Metallic crystals have spherical atoms that are held together, and they are bonded to each other in an equal fashion. A model can be created by packing spheres as close to each other as possible and using the appropriate amount of space. This is considered as closest packing. Spheres are stacked in layers where each sphere is surrounded by six others. The second layer of spheres occupy an indentation formed by three spheres over the first layer. The third layer occupies an indentation of the second layer in either of two ways. They can either occupy positions so that the spheres in the third layer are directly over a sphere in the first layer. This is considered to be the ABA arrangement. They can also occupy positions in the third layer in which none of the spheres lie directly on top of the first layer. This is considered to be the ABC arrangement.

An ABA arrangement has a hexagonal unit cell. This is considered to be hexagonal closest packed structure. An ABC arrangement is considered to be face-centered cubic unit cell. This is considered to be cubic closest packed structure. In the hexagonal structure, the same spheres occupy every other layer in a vertical position. In the cubic closest packed structure, the same spheres occupy every fourth layer in a vertical position. Both structures have each sphere with twelve nearest neighbors. There are six in the same layer, three in the layer above, and three in the layer below. It is important to know the net number of spheres for a specific unit cell. Let us take a look at a face-centered cubic unit cell. This type of unit cell has the centers of the spheres on cube's corners. There are eight cubes that share a sphere. There is one-eighth of the sphere that is inside each unit cell. If we take the product of eight times one-eighth, then we have one whole sphere. Spheres at the center of each face is shared by two unit cells. Therefore, there is one-half that lies in a specific unit cell. The cube itself has six faces, so there are six times one-half parts to make three whole spheres. The final number of spheres in a face-centered cubic unit cell is $(8 \times 1/8) + (6 \times 1/2) = 4$. Metals such as aluminum, iron, copper, cobalt, and nickel make cubic closest packed solids. The elements magnesium and zinc form hexagonal closest packed structures. Calcium and other kinds of metals crystallize in either structure. The alkali metals assume a structure type called body-centered cubic unit cell. This is a structure where the spheres are touching along the body diagonal of a cube. Each of the spheres have eight nearest neighbors. There are twelve nearest neighbors in the closest packed structures.

Metals: Bonding

Bonding models for metals gives an indication of the their physical properties. These include malleability, ductility, and conduction of heat and electricity in all directions. The shapes of metals can be changed easily. They are durable and have high melting points. Bonding for metals is strong and nondirectional. It is not easy to separate metal atoms, but it is easy to move them given the fact that the atoms stay close to each other. This can be explained based on the Electron Sea Model. The Electron Sea Model shows metal cations in a "sea" of valence electrons. These mobile electrons conduct both heat and electricity. The metal ions can be moved around easily as the metal is made into a sheet or pulled into a wire.

The band model or molecular orbital model gives the electron energies and motions. The electrons are assumed to travel around a metal crystal based on the formation of molecular orbitals from atomic orbitals. When there are many metal atoms that interact with each other, a lot of molecular orbitals become closely spaced and form a continuum of levels. These levels are called bands. Empty molecular orbitals that are close in energy to filled molecular orbitals provides an explanation of the thermal and electrical conductivity of metals. The reason why metals conduct electricity and heat is due to the highly mobile electrons. In the band model, mobile electrons take place when electrons in the filled molecular orbitals become excited into empty ones. The electrons travel throughout the metal crystal due to a potential on the metal. Molecular orbitals that have the conduction electrons are called conduction bands. These electrons that are mobile account for the conduction of heat through metals.

Alloys

Other elements can be put into a metallic crystal to make substances called alloys. An alloy is a substance that contains a mixture of elements. It also has properties characteristic of a metal. There are two types of alloys. Substitutional alloys have some of their metal atoms replaced by other metal atoms that are similar in size. Interstitial alloys have small atoms occupied in the interstices in the closest packed metal structure. Brass is an example of a substituted alloy. Some zinc atoms are replaced by copper atoms. Steel is an example as an interstitial alloy. Iron atoms surround carbon atoms that are occupied in the interstices. The physical and chemical properties of both alloys change. Alloys are stronger and harder than the metal itself. The element carbon affects the strength of steel. Mild steels contain less than 0.2% carbon. They are ductile and malleable. They are used for nails, cables, and chains. Medium steels contain 0.2 to 0.6% carbon. They are harder than mild steels. They are used in rails and steel beams. High-carbon steels have 0.6 to 1.5% carbon. They are tough and hard. They are used for springs, tools, and cutlery. There are other kinds of steel that contain elements besides iron and carbon. We call these alloy steels. They are mixed with interstitial and substitutional alloys.

Network Atomic Solids

Network solids are considered to be atomic solids that have directional covalent bonds. These materials are brittle, and they don't conduct electricity efficiently. Carbon and silicon are two elements. Diamond and graphite are two forms of carbon, and they are network solids. Diamond is considered to be the hardest substance in nature. Each carbon is surrounded by an arrangement that is tetrahedral and forms a big molecule. Covalent bonds stabilize the structure, and they are formed by overlapping sp^3 hybridized carbon atomic orbitals. This is when we look at the Localized Electron Model. Let us take a look at the Molecular Orbital Model. There is a large gap between filled and empty levels in an energy-level diagram for diamond so electrons can't be transferred to empty conduction bands. Therefore, diamond is not a good electrical conductor. We can say that diamond is an electrical insulator. Insulators don't conduct electric currents.

Graphite is a different form of carbon. It is slippery, and it is a conductor. This has to do with differences in bonding between the two structures. Graphite is envisioned as layers of carbon atoms that are in six-membered rings. The carbon atoms in a layer of graphite is surrounded by three other carbon atoms that is in a trigonal planar arrangement with 120 degree bond angles. The sp^2 orbitals form sigma bonds with three other carbon atoms. An unhybridized 2p orbital is perpendicular to the plane of carbon atoms. Pi molecular orbitals are formed. This gives stability for the layers. The delocalized electrons account for its electrical conductivity. The orbitals are spaced closely to each other and are similar to conduction bands in metal crystals. Graphite is slippery because of the strong bonding within layers of carbon atoms. But there is little to no bonding between the layer. This kind of arrangement allows the layers to pass one another easily. Diamonds are hard and because of this they are used for cutting. Graphite can be converted to diamond. Applying a pressure over 150,000 atm at around 2800 degrees Celsius converts graphite to diamond. Bonds are broken and rearrangement occurs.

Silicon makes up a good portion of the Earth's crust. Silicon compounds are formed from most rocks, sands, and soil that is formed on Earth's crust. Silicon compounds involve silicon-oxygen bonds that form chains. Silica is the basic unit for the silicon-oxygen compound. It has the empirical formula SiO_2. Quartz and sand are made up of silica. Silicon obeys the octet rule by making single bonds with four oxygen atoms. A network of SiO_4 that is in a tetrahedral arrangement is formed. Silicates are also found in rocks, soil, and clay. The oxygen to silicon ratio is greater than 2:1. They also have silicon-oxygen anions. Cations are also used to balance the charge. Silicates are considered to be salts that have metal cations and anions made up of silicon-oxygen. There are several different kinds of silicon anions. They have interesting and unusual shapes found in nature. If we heat silica above its melting point and cool it quickly, an amorphous kind of solid results. The solid that is made is called glass. Glass has a lot of disorder to it. Compounds such as Na_2CO_3 can be added to silica melt which is then cooled down. If B_2O_3 is added, then we have borosilicate glass. This is considered to be Pyrex.

Ceramics

Ceramics are made up of clays that is hardened using high temperatures. They are considered to be made up of nonmetallic elements that have the following characteristics. These include strong, brittleness, and resistance to heat and chemicals. Ceramics are made up of silicates. Ceramics are considered to be heterogeneous. Ceramics consist of silicates made in a glassy cement. Clays are made of water and carbon dioxide on feldspar. Feldspar consist of mixture of silicates having the empirical formula $K_2O*Al_2O_3*6SiO_2$ and $Na_2O*Al_2O_3*6SiO_2$. Feldspar is an aluminosilicate in which aluminum and silicon atoms are part of the polyanion that contains oxygen. When feldspar weathers, it produces kaolinite. Kaolinite is made of thin platelets that have the formula $Al_2Si_2O_5(OH)_4$. When the platelets become dry, they come together. If water is added, they slide over one another. This is why clay becomes like a plastic. When the clay dries, the platelets come together again. Firing the clay drives the water off which leads to silicates and cations forming glass. This binds tiny crystals of kaolinite. Ceramics are known for their stability being at high temperatures and its ability to resist corrosion. It is used to make jet and automobile engines. They have good fuel efficiency at high temperatures. However, there is a drawback using ceramics. They are brittle. Ceramics can be made flexible by adding a small amount of organic polymer. This is less subjected to fracture. This should be useful for lighter and long lasting engine parts, superconducting wires, and other small electronic devices. They can also be used for prosthetic devices like artificial bones.

Semiconductors

The energy gap for silicon is smaller. Some of the electrons cross the gap near room temperature. This makes silicon a semiconducting element or semiconductor. When the temperature is high, there is more energy to excite electrons into conduction bands. The ability of silicon to conduct is higher. This is a characteristic of a semiconducting element. Metals are different. Conductivity decreases for metals when the temperature increases. The conductivity of silicon can be greater if the silicon crystal is doped with other elements. If arsenic atoms replace some of the silicon atoms, there are extra electrons for conduction. Recall that there is more valence electrons for arsenic atoms than silicon atoms. This is called an n-type semiconductor. The extra electrons are close in energy to the conduction bands and are readily excited in their levels. This is where they can conduct an electric current.

The conductivity of silicon can also be greater if the element boron replaces some of the silicon atoms. Boron has one less electron than silicon. This creates a hole or an electron vacancy. When an electron fills this hole, a new hole is formed. The process is repeated several times. There is one electron in some of the molecular orbitals. The unpaired electrons can act as conducting electrons.

Thus, the substances can conduct a lot better. This is considered to be p-type semiconductors. These positive holes are thought of as charge carriers.

Applications involving semiconductors involve a p-type and an n-type to form a p-n junction. A junction is where a small number of electrons go from the n-type region into a p-type region. The p-type region is where there are holes in low energy molecular orbitals. A negative charge is placed on the p-type region and a positive charge on the n-type region. A charge is built up, and this is considered to be the contact potential or junction potential. This holds back electrons migrating. If a negative terminal of a battery is connected to the p-type region and the positive terminal of the battery is connected to the n-type region, electrons are pulled towards the positive terminal. There are then holes in the negative terminal. This is opposite to the flow of electrons at a p-n junction. The junction resist current to flow, and this is said to be in reverse bias. There is no current flowing in this system. If a negative terminal is connected to an n-type region and a positive terminal is connected to a p-type region, the movement of electrons and holes migrating is in the desired direction. This junction has current flowing easily. This is considered to be forward bias. A p-n junction is used as a rectifier. Rectifiers are devices that make a current flow in one direction from alternating currents. If a rectifier is put into a circuit where the potential is in reverse, a p-n junction gives current under forward bias. This converts alternating current to direct current. The p-n junction is used in electronics.

Molecular Solids

There are solids where the atoms occupy lattice positions. Network solids can be thought of as one giant molecule. There are also solids that contain discrete molecular units that occupy lattice positions. Ice, dry ice which is solid carbon dioxide, and S_8 molecules as well as P_4 molecules are considered to be molecular solids. The substances have strong covalent bonding and have weak forces between the molecules. Forces among the molecules in a molecular solid has to do with the nature of the molecule. Molecules such as CO_2, I_2, P_4, and S_8 do not have a dipole moment. The kind of intermolecular forces they have are London Dispersion forces. The forces are small and the substances are in their gaseous state. The size of the molecule increases. London Dispersion forces then become large which causes these substances to be solids near room temperature. When molecules have dipole moments, they have bigger intermolecular forces. This is especially true when there is hydrogen bonding forces present.

Ionic Solids

Ionic solids are considered to be stable substances that have high melting points. Cation and anions are held together by strong electrostatic forces that occur between oppositely charged ions. Let us take a look at the closest packing of spheres. The anions are placed in the closest packing arrangement either as hexagonal closest packed or cubic closest packed. The small cations fill in the holes among the closely spaced anions. The ions are packed in a way where there is maximum electrostatic attraction between oppositely charged ions, and there is a minimum repulsion of the ions with like charges. Let us look at three types of holes in closest packed structures. Trigonal holes are obtained when three spheres are in the same plane. Tetrahedral holes are obtained when one sphere sits on the three spheres in the same plane. Octahedral holes are obtained when three spheres formed in a triangle are on top of three spheres that are in a triangle, but the triangles are pointing away from each other. The holes get bigger from trigonal to tetrahedral to octahedral. Trigonal holes are small, and they are not occupied in binary ionic compounds. The size of the anion and cation depend on the tetrahedral or octahedral holes in a binary ionic solid. Closest packed structures have twice as many tetrahedral holes as spheres packed together. Closest packed structures have the same number of octahedral holes as spheres packed together. Ions are considered to be hard spheres maximizing attractions and minimizing repulsions

Vapor Pressure

Let us take a look at a change of state. A liquid evaporating from an open container is an example of a change in state. Molecules of a liquid escape the liquid's surface and make a gas. This is considered to be vaporization or evaporation. Vaporization is considered to be endothermic. This means that energy is used to break apart intermolecular forces in the liquid. The energy that is required to vaporize 1 mole of a liquid at atmosphere is considered as the heat of vaporization. It is also called the enthalpy of vaporization. It has the symbol ΔH_{vap}. Water acts as a coolant. Water has strong hydrogen bonding in its liquid state. It also has a large heat of vaporization. A lot of the sun's energy is used to evaporate water from oceans, lakes, and rivers. Water is also important for the body's temperature by evaporation of perspiration.

Suppose a liquid is put into a closed container. The amount of liquid decreases and becomes constant. Molecules from the liquid phase enter the vapor phase. Evaporation happens at a constant rate at a certain temperature. But let us take a look at the reverse process. If the number of vapor molecules increase, the rate of return of these molecules also increases. Condensation takes place. Condensation can be considered to be when vapor molecules form into a liquid. The rate of condensation then equals the rate of evaporation. There is no further change that takes place and the processes are exactly balanced. This is when the system is at equilibrium. The process is dynamic at the molecular level. The opposite processes balance each other, and there is no net charge.

The pressure of the vapor that is at equilibrium is considered to be the vapor pressure of the liquid. Liquids that have high vapor pressures is considered to be volatile. In other words, they evaporate rapidly. The vapor pressure is determined by the size of the intermolecular forces. Large intermolecular forces have low vapor pressures. The molecules need a lot of energy to enter into the vapor phase. Substances that have large molar masses have low vapor pressures. This is due to large dispersion forces. If there are more electrons, the more polarizable the substance is. That means there are greater dispersion forces. Vapor pressure increase with temperature. For a molecule to overcome its intermolecular forces, the molecule must have a certain amount of kinetic energy. As the temperature increases, the molecules needed to overcome these forces increasingly enter the vapor phase. The equation for vapor pressure and temperature would be as follows

$$\ln(P_{vap}) = -\Delta H_{vap}/R(1/T) + C$$

where ΔH_{vap} is the enthalpy of vaporization, R is a gas constant, and C is a constant. The following equation is called the Clausius-Clapeyron equation.

$$\ln(P_{vap1}/P_{vap2}) = \Delta H_{vap}/R(1/T_2 - 1/T_1)$$

Solids have vapor pressures too. Iodine placed in a beaker being heated and an evaporating dish filled with ice placed on top of the beaker gives vapor. The vapor is in equilibrium with the solid. In other words, iodine sublimes from the solid state to the vapor state. Sublimation takes place from the solid state directly to the gaseous state. Solid carbon dioxide which is also known as dry ice goes through sublimation.

Changes of Different States

If a solid is heated, it will melt and a liquid will be formed. Continue heating the sample will cause the liquid to boil and form into the vapor phase. This is described as a heating curve. Time versus temperature is plotted on a graph. Energy is added at a constant rate. Let us take a look at solid ice. As energy goes into ice, there are random vibrations for the water molecules. The vibrations increase as the temperature continues to rise. The molecules break apart from their lattice positions. Solids change their state to liquids. At zero degrees celsius which is the melting point of ice, solid changes to liquid where energy is being added by breaking the hydrogen bonds of the ice structure. The potential energy of the water molecules increase. The enthalpy of fusion takes place at the melting point when the solid is melting. The temperature stays constant until the solid is changed to a liquid. The temperature then increases. When the temperature reaches one hundred degrees celsius, this is considered to be the boiling point. The temperature then stays the same as energy is added to vaporize the liquid. As all the liquid is changed to vapor, the temperature rises again. These are all physical changes of state. However, if water vapor is heated even at higher temperatures, water will break apart into hydrogen and oxygen gas. This would be a chemical change.

Melting and boiling points are determined by vapor pressures of solid and liquid states. If we consider the temperature below zero degrees celsius, we find that the vapor pressure of ice is smaller than the vapor pressure of liquid water. The vapor pressure of ice has a larger temperature dependence than liquid water. The vapor pressure of ice increases a lot more rapidly than the vapor pressure of water. As the temperature of the water increases, there comes a point when the liquid and solid have identical vapor pressures. This is called the melting point. If the vapor pressure of the solid is greater than the liquid, vapor is released from the solid and the liquid will absorb vapor to reach equilibrium.

Only the liquid state exists above the melting point of ice. If the vapor pressure of the solid is less than that of the liquid, the liquid slowly disappears, and the amount of ice gets bigger. Solid remains and equilibrium is achieved with its vapor. The solid state exists only at the melting point of ice. If the vapor pressures of the solid and liquid are the same, both of them are at equilibrium. This represents the freezing point. Both the solid and the liquid states exist. The normal melting point is the temperature which the solid and liquid states have the same vapor pressure when the total pressure is 1 atmosphere. The normal boiling point is the temperature which the vapor pressure of the liquid is exactly at 1 atmosphere. Changes of state do not always happen at the boiling or melting point. Water can be supercooled. This means that it can be cooled below zero degrees celsius at 1 atm, and it should stay in the liquid state.

Supercooling happens because as water is cooled, the water molecules do not achieve the degree of organization to make ice at zero degrees celsius. It stays as a liquid. Ordering takes place and ice then forms. Energy is released in the exothermic process and temperature is brought back up to the melting point. This is where the rest of the water freezes. Liquids can also be superheated. This means that it can be raised to temperatures above its boiling point. Bubbles formed in the interior of the liquid require high-energy molecules to gather near the same vicinity. This might not happen at the boiling point if the liquid is heated off fast. The vapor pressure in the liquid becomes greater than the atmospheric pressure. The bubble burst before rising to the surface and blowing liquid out of the container. This is considered to be bumping. Boiling chips added to the flask can solve this problem. Boiling chips are small pieces of porous ceramic material that contains trapped air and escapes upon heating. Tiny bubbles are then formed. A smooth flow of boiling occurs when the boiling point is reached.

Phase Diagrams

Phase diagrams are diagrams that represent the phases of a substance as a function of temperature and pressure. Phase diagrams describe conditions on a closed system. Materials cannot escape into the surroundings. Air is not present. Phase diagrams are graphed as temperature in degrees celsius on the x-axis and pressure in atmospheres on the y-axis. Three phases are shown on the phase diagram. These are solid, liquid, and gas. The triple point is where all three phases exist.

Conclusion

Liquids and solids are condensed states of matter. Liquids and solids are formed by intermolecular forces among molecules, atoms, or ions. Dipole-dipole forces is a type of attraction between molecules that have dipole moments. The forces are strong for molecules that contain hydrogen bonded to an electronegative element such as nitrogen, oxygen, or fluorine. These kinds of dipole forces are called hydrogen bonds. They have very high boiling points. London Dispersion forces are caused by instantaneous dipoles. Nonpolar molecules have London Dispersion forces. Liquids also have many properties such as surface tension, capillary action, and viscosity. They depend on the strengths of intermolecular forces.

Solids are also classified as either crystalline or amorphous. Crystalline solids are arranged regularly, and its smallest repeating unit is the unit cell. Crystalline solids are classified to the type of components that occupy the lattice points such as atomic, ionic, or molecular solids. Particles in a crystalline solid is determined by X-ray diffraction. The interatomic distances is related to the wavelength and angle of reflection of X-rays by the Bragg Equation. Metals can be structured by assuming atoms to be spheres that compact themselves as possible. Two closest packing arrangements are hexagonal and cubic. Bonding in metallic crystals is described by the Electron Sea Model or the Band Model. The Band Model consists of a large number of atomic orbitals that form closely spaced energy levels. Electricity is conducted by electrons in conduction bands. These conduction bands are molecular orbitals that have only one electron. Metals form two types of alloys, and they can be classified as substitutional or interstitial. Carbon is a network solid that contains strong directional covalent bonds. Diamond and graphite are allotropes of carbon, and they have different physical properties based on different bonding.

Silicon compounds contain a network of interconnected SiO_4 tetrahedra. Silicate salts contain polyatomic anions of silicon and oxygen. They are components of rocks, soil, clays, glass, and ceramics. Semiconductors are formed when silicon gets doped with other elements. An n-type semiconductor have atoms with more than valence electrons than the silicon atom. A p-type semiconductor have atoms with fewer valence electrons than silicon. The electronics industry is based on devices with p-n junctions. Molecular solids have small molecules at lattice points. They are held together by weak intermolecular forces. An example is ice. Ionic solids, however, are held together by electrostatic attractions. Crystal structure of ionic solids are described by fitting cations into holes into closest packed structures formed by anions.

A phase change that takes place between a liquid and vapor is called evaporation. The heat of vaporization is the required energy to change from one mole of liquid to its vapor. Condensation is a reverse process. This is where vapor molecules return to their liquid state. The balance of the rate of evaporation and condensation in a closed container is considered for the system to be in equilibrium. The pressure of the vapor is referred to as vapor pressure. The volatility of liquids depend on intermolecular forces. Liquids that have strong intermolecular forces have low vapor pressures. The normal melting point of a solid is the temperature at which the solid and liquid have identical vapor pressures. The normal boiling point of a liquid happens at a temperature where the vapor pressure of the liquid is 1 atmosphere. A phase diagram tells us which state exists at a temperature and pressure. The triple point tells us the temperature and pressure where three states coexist. The critical point is the point where the critical temperature and critical pressure exist. Critical temperature is the temperature above which the vapor cannot be liquefied no matter what pressure is applied. The critical pressure is the pressure needed to make liquefaction at the critical temperature.

Chapter 12

Solution Chemistry

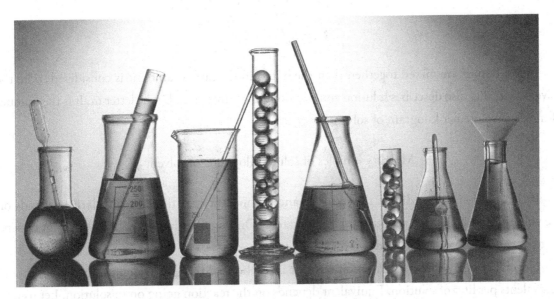

Test-Tubes with Blue Liquid on Blue Background

Most solutions are mixtures. Examples include wood, milk, seawater, and air. If the mixture is uniformly mixed, it is called a homogenous mixture. Homogeneous mixtures are also called solutions. Solutions can be either gases, liquids, or solids. Liquid solutions especially containing water are important. A lot of chemical reactions take place in aqueous solutions. This is due to the fact that water is able to dissolve many substances.

The Composition of Solution

Mixtures have variable composition. The amount of substances has to be specified. If very little solute is present, the solution is considered dilute. If a large amount of solute is present, the solution is considered concentrated. Solution composition should be defined more precisely in order to carry out the calculations. Solution composition can be expressed in terms of molarity. Molarity is defined as the number of moles of solute per liter of solution. It is symbolized by the capital letter M. Solutes are substances that are being dissolved and the solvent is the dissolving medium. Mass percent is sometimes referred to as weight percent. It is the percentage of mass of solute in solution. Mass percent can be expressed as

Mass Percent = (mass of solute/mass of solution) X 100%

Solution composition can also be described as mole fraction. This is symbolized by the Greek lowercase letter chi, $\aleph$. It is the ratio of the total number of moles of a component to the total number of moles of solution. If the solution is two-component, and if we let n_A and n_B represent the number of moles, then we have the following equation.

$$\aleph_A = n_A/(n_A + n_B)$$

$$\aleph_B = n_B/(n_A + n_B)$$

If two liquids are mixed together, then the liquid in the largest amount is considered to be the solvent. Molality also describes solution composition. It is symbolized by the letter m. It is the number of moles of solute per kilogram of solvent.

Molality = moles of solute/kilograms of solvent

If the solution is very dilute, the molality and the molarity are the same. Molarity depends on the volume, and it changes slightly with temperature. On the other hand, molality does not depend on temperature. Molality depends on mass.

Normality is another concentration term. It is symbolized by N. It is defined as the number of equivalents per liter of solution. Equivalent depends on the reaction going on in solution. Let us take a look at an acid-base reaction. An equivalent is the mass of acid or base that can give or accept 1 mole of H^+ ions. For sulfuric acid, the equivalent mass is the molar mass divided by 2. This is true because each mole of sulfuric acid gives 2 moles of protons. An equivalent can be defined as 1 equivalent of acid that reacts with 1 equivalent of base. Let us take a look at oxidation-reduction reactions. An equivalent is the quantity of oxidizing or reducing agent that accepts or gives 1 mole of electrons. One equivalent of a reducing agent reacts with 1 equivalent of an oxidizing agent. The equivalent mass of an oxidizing or reducing agent can be found by the number of electrons in a half-reaction.

Energies involving Solutions

Solutes being dissolved in solution is common. For example, salt is dissolved in water to cook food, carbon dioxide in its gaseous state in water is used to make soda, and ethanol used in gasoline is used to make gasohol, etc. Solubility is also important. There are factors that affect solubility. Solubility can be thought of as "like dissolves like". Polar or ionic solutes dissolve in polar solvents. Nonpolar solutes dissolve in non polar solvents. A liquid forming in solution takes place in three steps. The first step is separating the solute into individual components. The next step is to overcome intermolecular forces in the solvent so that there is space for the solute. The last step is to allow solute and solvent to interact with each other and form a solution. In other words, the solute and solvent are expanded and then are combined together to form a solution. The first two steps require energy. Intermolecular forces must be overcome for the solute and solvent to expand. The last step releases energy. The first two steps are endothermic, and the last step is exothermic. The enthalpy or heat of solution is the enthalpy change when the solution forms. It is symbolized by ΔH_{soln}.

$$\Delta H_{soln} = \Delta H_1 + \Delta H_2 + \Delta H_3$$

The enthalpy of solution can also be defined as the sum of the energies that is used to expand both solvent and solute and the energy of solvent-solute interactions. The symbol ΔH_{soln} can be positive which means energy is absorbed or it can be negative which means energy is released. The enthalpy of hydration, which is also known as the heat of hydration, combines two terms. The ΔH_2 and ΔH_3 terms are combined. The symbol for the enthalpy or heat of hydration is ΔH_{hyd}. The ΔH_2 is used for expanding the solvent, and the term ΔH_3 is used for solvent-solute interactions. The heat of hydration is the enthalpy change when gaseous solute is dispersed in water. A process is favored by an increase in probability. Energy considerations are also important. Large amounts of energy are required for processes that do not occur. Solutions form because there is a higher chance of probability of a mixed state.

Solubility

Solubility is favored when the solute and solvent have polarities that are the same. Molecular structure determines polarity. There should be a connection between structure and solubility. Let us take a look at the structure of vitamins. Vitamins can be divided into two classes. Fat-soluble vitamins include Vitamins A, D, E, and K. Water-soluble vitamins include Vitamins B and C. Vitamin C contains more polar bonds than Vitamin A. Vitamin C tends to be water-soluble, whereas Vitamin A tends to be water insoluble. Vitamin A is water insoluble because it is nonpolar. It tends to be soluble in nonpolar materials like body fat. Vitamin A can also be thought of as being hydrophobic which means water-fearing. Vitamin C, on the other hand, can be thought as being hydrophilic which means water-loving. Fat-soluble vitamins build up in the fatty tissues of the body. Water-soluble vitamins get excreted by the body and have to be consumed regularly. You can see how important solubility is when we study the structure of each of the molecules. Each structure is unique in based on their chemical and physical properties. Nevertheless, solubility allows the scientist to separate polar and nonpolar molecules for further studies.

Effects of Pressure

Pressure have little effects on solubilities of solids or liquids. However, pressure does increase the solubility of a gas. Henry's Law states there is a relationship between gas pressure and concentration of dissolved gas. The equation for Henry's Law is

$$C = k \times P$$

where C stands for the concentration of dissolved gas, k is a constant for a particular solution, and P represents the partial pressure of the solute that is gaseous above the solution. Henry's Law states the amount of gas dissolved in a solution is directly proportional to the pressure of a gas that is above a solution. Henry's Law works best for dilute solutions of gases that do not react with the solvent.

Effects of Temperature

Solubility for most substances increase with temperature. Some ionic compounds such as sodium sulfate and cerium sulfate decrease solubility with increasing temperature. Dissolving of a solid occurs really fast at higher temperatures. Solids can increase or decrease with temperature. There is some relationship between the sign of ΔH°_{soln} and solubility with temperature. There are still important exceptions. The only way to determine how temperature depends on solubility is by doing the experiment. Gases dissolving in water is not that complicated. The solubility of a gas in water tends to decrease with increasing temperature. A temperature effect like this has important environmental applications due to the widespread use of water from lakes and rivers for industrial cooking. After water is used, it is returned to a natural source at a higher temperature. This is when thermal pollution has occurred. The water then becomes polluted water and contains less than normal concentration of oxygen. It is also less dense. The warmer water floats on the colder water blocking oxygen being consumed. This effect has consequences for deep lakes. A warmer upper layer of water decreases how much oxygen is available to aquatic species in deep layers of the lake.

Gases that decrease solubility when temperature increases is responsible for the formation of a boiler scale. The reaction of a carbonate ion, carbon dioxide, and water leads to the formation of bicarbonate. If there are calcium cations around, we find that calcium carbonate is soluble in water. However, calcium carbonate is insoluble. When water is heated, carbon dioxide is released. In order for carbon dioxide to be replaced, the reverse reaction of bicarbonate should occur. That is, bicarbonate decomposes into water, carbon dioxide, and the carbonate anion. The reaction increases the concentration of carbonate ions. Calcium carbonate then forms. We can think of the solid as the boiler scale that forms on the walls of the container like tea kettles. Boiler scale reduces the transfer of heat and leads to pipes being blocked.

Vapor Pressures of Solutions

The physical properties of liquid solutions are different from the pure solvent itself. Nonvolatile solutes lower the vapor pressure of the solvent. Nonvolatile solutes do not have any tendency to escape from the solution and enter the vapor phase. Studies of vapor pressures of solutions were carried out by Francois M. Raoult. He came up with the following equation known as Raoult's Law

$$P_{soln} = X_{solvent} P^{o}_{solvent}$$

where P_{soln} is the vapor pressure that is observed for the solution, $X_{solvent}$ is the mole fraction of solvent, and $P^{o}_{solvent}$ is the vapor pressure of pure solvent. The nonvolatile solute basically dilutes the solvent. Raoult's Law states the vapor pressure of a solution is directly proportional to the mole fraction of solvent.

Lowering vapor pressure helps us determine how many molecules there are and for determining molar masses. If we dissolve a mass of the compound in a solvent, the vapor pressure of the solution is measured. Using the equation of Raoult's Law, the number of moles of solute can be determined. The molar mass can then be calculated by taking the ratio of the mass to moles. Lowering vapor pressure depends on the number of solute particles in the solution. Vapor pressure measurements is also used to characterize solutions. If a salt like potassium chloride is dissolved in water, the vapor pressure is lowered twice as expected. This happens because the solid has two ions per formula unit. The ions separate and dissolve. Vapor pressure measurements provide information about the solute after dissolving in solution.

Nonideal Solutions

At this point, we assumed the solute is nonvolatile and does not contribute to the vapor pressure over the solution. Liquid-liquid solutions where both components are volatile, the equation for Raoult's Law is modified. The equation takes the form

$$P_{Total} = P_A + P_B = X_A P_A^{\,o} + X_B P_B^{\,o}$$

where P_{Total} is the total vapor pressure of the solution for components A and B, X_A and X_B are mole fractions for A and B, $P_A^{\,o}$ and $P_B^{\,o}$ are vapor pressures of the pure components A and B, and P_A and P_B are partial pressures for molecules in the vapor above the solution. Liquid-liquid solutions that obey Raoult's Law are called ideal solutions. The ideal behavior of solutions is not perfect, but it is closely approached. Ideal behavior normally happens when solute-solute, solvent-solvent, and solute-solvent interactions are the same. Solutes dilute solvents in solution if the solute and solvent are close to each other. If the solvent has an attraction for the solute such as hydrogen bonding, the chances of the solvent molecules escaping is small. The observed vapor pressure will be smaller than what Raoult's Law predicted. A negative deviation of Raoult's Law results.

If the solute and solvent release a huge quantity of energy forming a solution or when ΔH_{soln} is large and negative, their must be strong interactions between solute and solvent. A negative deviation results from Raoult's Law since both of the components have a smaller chance of escaping in solution than in pure liquids. An example of this would be acetone-water where molecules hydrogen bond with each other. Suppose that two liquids mix endothermically, this tells us that solute-solvent interactions are weaker than interactions among molecules in pure liquids. It takes a lot more energy to expand liquids to be released when liquids are mixed. Molecules in solution have a better chance to escape than what we expect. Positive deviations of Raoult's Law are observed. Let us take a look at propanol and pentane. Proponal is polar and pentane is nonpolar. They do not interact effectively. The enthalpy of solution becomes positive. For solutions of similar liquids like benzene and toluene, the enthalpy of solution is close to zero. The solution eventually obeys Raoult's Law which is close to ideal behavior.

Boiling Point Elevation and Freezing-Point Depression

At this point, we learned that a solute affects the vapor pressure of a liquid solvent. Changing one state to another depends on vapor pressure. Adding a solute affects the freezing and boiling point of a solvent. Collectively, freezing-point depression, boiling-point elevation, and osmotic pressure are called colligative properties. They depend on the number and not the identity of solute particles in an ideal solution. They have a direct relationship to the number of solute particles. Colligative properties are helpful for determining the nature of a solute after being dissolved in a solvent and determining molar masses of substances. The normal boiling point for a liquid is when the vapor pressure is equal to 1 atmosphere. Nonvolatile solutes lower the vapor pressure of a solvent. The solution has to be heated to a higher temperature than the boiling point of a pure solvent. The solution has to be heated to a higher temperature than the boiling point of a pure solvent. This has to happen because the vapor pressure has to reach 1 atmosphere. A nonvolatile solute elevates the boiling point for a solvent. Boiling-point elevation depends on the concentration of solute. The equation for the change in boiling point can be represented as

$$\Delta T = K_b \times m_{solute}$$

where ΔT is the boiling-point elevation or the difference between the boiling point of a solution and that of a pure solvent, K_b, is a constant that has to do with the solvent and is called the molal boiling-point elevation constant and m_{solute} is molality of solute in solution. Nonvolatile solutes extend the liquid range of solvents.

If a solute is dissolved in a solvent, the solution's freezing point is lower than the pure solvent. The vapor pressures of ice and liquid water should be the same at zero degrees Celsius. If a solute is dissolved in water, the solution does not freeze at zero degrees Celsius. Water has a lower vapor pressure than ice. Ice, in this case, does not form. Note the vapor pressure of ice tends to decrease faster than liquid water when the temperature goes down. Cooling the solution allows the vapor pressure of ice and liquid water to become equal. The temperature ends up being smaller than zero degrees Celsius. This is the new freezing point of the solution. The freezing point becomes depressed. Adding solute to the liquid lowers the rate molecules go back to its solid state. Liquid states are formed at zero degrees Celsius for an aqueous solution. Cooling the solution allows the water molecules to form solid ice. Equilibrium is reached. We consider this the freezing point of the water in solution. Solutes lower the freezing point of water. The equation for freezing-point depression is

$$\Delta T = K_f \times m_{solute}$$

where ΔT is the freezing-point depression or the difference between the freezing point of pure solvent and that of solution, K_f is a constant for the solvent, and it is called the molal freezing-point depression constant. The freezing-point depression is used to determine molar masses and tell us something about the solution.

Osmotic Pressure

Solutions and solvents are separated by a semipermeable membrane. Solvent molecules but not solute molecules pass through the semipermeable membrane. Eventually, the volume of the solution increases and the volume of the solvent decreases. The flow of solvent into the solution through the semipermeable membrane is a process called osmosis. Sooner or later, the level of the liquids stop changing. This is where equilibrium occurs. The liquid levels are not the same at this point. There is a larger hydrostatic pressure on the solution than the pure solvent. The osmotic pressure is the excess pressure. Let us take a look at what else osmosis can do. We can prevent osmosis by applying pressure to the solution. The smallest pressure that stops the osmosis equals the osmotic pressure of the solution. The membrane allows solvent molecules to go through. The initial rates of solvent transfer from both sides is not the same. Solute particles interfere with solvent going through to the pure solvent side. There is a net transfer of solvent molecules going into the solution. The solution ends up having a higher volume. When the solution rises in the U-tube, the pressure gives a stronger push on the solvent molecules in the solution. This forces them back to the membrane. Enough pressure has been developed so the transfer of solvents is equal in both directions. Equilibrium has been reached and the levels do not change. We can use osmotic pressure to study solutions and determine molar mass. Osmotic pressure is useful. Small concentration of solutes produce large osmotic pressure. The equation for osmotic pressure is

$$\Pi = MRT$$

where Π is the osmotic pressure in atmospheres, M is the molarity of solution, R is the gas law constant, and T is the Kelvin temperature. Osmotic pressure measurements give accurate molar mass values than freezing-point or boiling-point changes. During the process of osmosis, a semipermeable membrane makes sure not all solute particles are going through. This process is similar to dialysis. The process of dialysis happens at the walls of most plant and animal cells. The membrane allows solvent molecules, small solute molecules, and ions to pass through.

Artificial kidney machines is an application of dialysis where blood is purified. Blood goes through the cellophane tube. The cellophane tube acts as a semipermeable membrane. The tube is placed in a dialyzing solution. The solution has the same concentration of ions and small molecules as regular blood. But there are waste products removed by the kidneys. The movement of these waste molecules into the washing solution makes the blood clean. Isotonic solutions have identical osmotic pressures. Fluids that are given intravenously should be isotonic with body fluids. If we put red blood cells in a hypertonic solution, the cells shrivel due to a net transfer of water out of the cells. The osmotic pressure in a hypertonic solution is higher than that of cell fluids. When water leaves the cells as described in the above process, this process is called crenation. Hemolysis is the opposite process. If red blood cells are bathed in a hypotonic solution, then hydrolysis occurs. The osmotic pressure in a hypotonic solution is lower than the cell fluids. The cell breaks open because water is flowing into the cells.

Reverse osmosis takes place when there is a net flow of solvent from a solution to a solvent that is separated by a semipermeable membrane. This happens when the external pressure is larger than the osmotic pressure. The semipermeable membrane described here is like a filter that removes solute particles. An application of this is the desalination of seawater. Seawater is found to be hypertonic to body fluids so we cannot drink seawater.

Colligative Properties

Colligative properties depend on the total concentration of solutes. The moles of solute dissolved and the moles of particles in solution is expressed using the van't Hoff factor, i

i = moles of particles in solution/moles of solute dissolved

The value for i can be found for a salt by counting the number of ions per formula unit. The salt sodium chloride has an i value of 2, whereas the salt Li_2CO_3 has an i value of 3. The values are chosen because we assume that if a salt dissolves, it dissociates completely into its component ions. The ions move around independently, but this is not always true. The experimental values, however, are close to the actual values. Sometimes, the experimental values are not the same as the calculated values. The reason for this is that ion pairing occurs in solution. In a small amount of time, the sodium and chloride ions are paired, and they count as a single particle. Ion pairing is important when the solution is concentrated. If the solution becomes more dilute, the ions are further apart, and there is a smaller chance that ion pairing occurs.

Ion pairing takes place to some extent in electrolyte solutions. The value i deviates more from the expected value when the ions have multiple charges. We expect this because ion pairing to be important for highly charged ions. The colligative properties of these electrolyte solutions include the van't Hoff factor. Changes in freezing and boiling points has the equation in modified form as

$$\Delta T = imK$$

where the symbol K is the freezing-point depression or boiling-point elevation constant for the solvent. The osmotic pressure of electrolyte solutions include

$$\Pi = iMRT$$

Colloids

If mud is mixed with water, we would have a solution of mud and water where most of the mud will settle down at the bottom. There will be some particles floating in solution. The presence of these small particles can be demonstrated by shining a beam of intense light through the suspension. The beam can be seen from the side of a glass container because light is scattered by the suspended particles. In a true solution, the beam is invisible from the side because ions and molecules dispersed in a solution is too small to scatter visible light. The Tyndall effect is the scattering of light by particles. It is used to tell the difference between a suspension and a true solution.

A suspension of small particles in a medium is called a colloidal dispersion or a colloid. The particles that are suspended are aggregates of molecules or ions. Colloids can be categorized based on the state of the dispersed phase and dispersing medium. The main factor that stabilizes a colloid is electrostatic repulsion. Colloids are electrically neutral. If a colloid is placed in an electric field, then the dispersed particles migrate to a same electrode and have the same charge. The center of a colloidal particle attracts ions with the same charge. These ions attract another layer of oppositely charged ions. Since colloidal particles have the same outer layer of ions with the same charge, they repel each other and do not aggregate to form particles that cannot penetrate.

Coagulation is a process of destroying the colloid. This is done by heating or adding an electrolyte. Heating increases the speed of colloidal particles. They collide with enough energy that ion barriers are penetrated and particles aggregate. The process is repeated so many times and the particle grows to a point where it settles into solution. If an electrolyte is added, then it neutralizes the adsorbed ion layers. Removing soot from smoke is also an example of coagulation of a colloid. If smoke is passed through an electrostatic precipitator, suspended solids are removed. Precipitators such as these has made an immense improvement in air quality.

Conclusion

We can describe solution composition based on molarity, mass percent, mole fraction, molality, or normality. The enthalpy of solution is a change for a solution to form. It can be described as either as an endothermic or exothermic process. There are also factors that affect solubility. A molecule's structure determines where it is polar or nonpolar. Polar molecules are soluble in polar solvents. Nonpolar molecules are soluble in nonpolar solvents. Pressure does not have a whole lot of effect where liquids or solids are soluble. Increasing pressure increases the solubility of gases. Henry's Law tells us that the concentration of a gaseous solute in a solvent is proportional to the partial pressure of the gas that is above the solution. Increasing temperature decreases the solubility of a gas in water. The solubility of a solid in water depends on the identity of the solid.

A nonvolatile solute's vapor pressure is less than the vapor pressure of a pure solvent. The solute decreases the number of solvent molecules per unit volume. This also lowers the solvent molecules from escaping. Raoult's Law tells us the vapor pressure of a solution is proportional to the mole fraction of the solvent. Raoult's Law also applies to ideal solutions. Nonideal behavior occurs if a solute has an affinity for a solvent. A negative deviation from Raoult's Law occurs. The vapor pressure is lowered than it is predicted. Colligative properties depend on the number of solute particles, boiling-point elevation, freezing-point depression, and osmotic pressure. Nonvolatile solutes lower the freezing point and raise the boiling point. It makes the liquid range of the solvent larger.

Osmosis takes place when a pure solvent and the solution are separated by a semipermeable membrane. This prevents the passage of solutes. Solvent flows from the solvent to the solution. Osmotic pressure is pressure that is applied to a solution that prevents osmosis. It is related to the molarity of a solution. Reverse osmosis happens when an external pressure that is larger than the solution's osmotic pressure is applied. There is a net flow of solvent from the solution into pure solvent. Colligative properties of electrolyte solutions are described by having the number of ions taken into account and by using the van't Hoff factor, i. The experimental value for i is less than the expected value when ion pairing occurs. Colloids are suspensions of small particles in solution. They are stabilized because of electrostatic repulsions between ion layers surrounding particles. Colloids undergo coagulation by either heating or the addition of an electrolyte.

Chapter 13

Kinetics of Reactions

Abstract Purple Crystal

A characteristic of a reaction is its spontaneity. Spontaneity is the tendency for a process to occur. But this implies nothing about its speed. Spontaneous has nothing to do with whether something is fast or slow. Useful reactions occur at a reasonable rate. It is important to study the stoichiometry, thermodynamics, and rate of a reaction. Chemical kinetics is the branch of chemistry that has to do with reaction rates. Chemical kinetics helps us understand the steps by which a reaction takes place.

The series of steps are called reaction mechanisms. Our understanding of mechanisms allows us to facilitate the reaction. Let us take a look at rate laws, reaction mechanisms, and models of chemical reactions.

Reactions Rates

Chemical kinetics deal with speed at which changes occur. The speed or rate is the change in a given quantity over a period of time. The quantity that changes is the concentration of reactant or product. The reaction rate is the change in concentration of a reactant or product per unit time.

$$Rate = \Delta[A]/\Delta t$$

where A is the reactant or product, and the square brackets indicate concentration in mol/L. The symbol Δ tells us a change in a given quantity. A change can be an increase or a decrease giving a positive or negative reaction rate. The value of a rate at a specified time is the instantaneous rate. It is obtained by computing the slope of a line tangent to a curve at a point. Since reaction rate changes with time and because rates are different, a clear specification of the rate of a reaction is important.

Chemical reactions are considered to be reversible. Studying a reaction after the reactants are mixed and before the products had time to build up to significant levels is important. If conditions are chosen where the reverse reaction is neglected, the reaction rate depends primarily on the concentration of the reactants. Let us take a look at a basic example. Consider the rate expression

$$Rate = k[A]^n$$

This expression shows that the rate depends on concentration of reactants known as the rate law. The proportionality constant, k, is normally referred to as the rate constant. The small letter, n, is called the order of the reactant. Both k and n is determined by experiment. The order in the rate expression can be an integer including zero or perhaps even a fraction. Keep in mind the rate expression above indicates the concentration of the products are not taken into account because the reverse reaction does not contribute to the overall rate. The order of the reaction is determined by experiment. Remember reaction rate means a change in concentration per unit time.

Rate laws that express rate and that depends on concentration is called the differential rate law. Sometimes, it is just called the rate law. Another kind of rate law is known as the integrated rate law. This expresses concentration depending on time. If we know the differential rate law, then we can determine the integrated rate law and vice versa. Determining the rate law for a reaction is important. We can work backward from the rate law equation and determine the steps the reaction occurs. Most chemical reactions take place in a series of sequential steps. If we are going to understand a chemical reaction, we need to understand what the steps are. A chemist is normally not interested too much in the rate law, but the steps by which a reaction occurs.

The Integrated Rate Law

We have expressed rate as a function of the reactant concentrations. The react ant concentrations can also be expressed as a function of time if the differential rate law is given for the reaction. Reactions involving one reactant can be written in the form

$$aA \longrightarrow Products$$

The rate law then takes the form

$$Rate = -\Delta[A]/\Delta t = k[A]^n$$

First-Order Rate Laws

Consider the reaction of the form

$$aA \longrightarrow Products$$

the rate law becomes

$$Rate = -\Delta[A]/\Delta t = k[A]^1$$

The integrated first-order rate law then becomes

$$\ln[A] = -kt + \ln[A]_o$$

The equation can also be expressed as

$$\ln([A]_o/[A]) = kt$$

First-Order Half Life

The half-life is the time for any reactant to reach one half of the original concentration. It is designated as the symbol $t_{1/2}$.

Consider the following derivation

$$\ln([A]_o/[A]) = kt$$

Let $t = t_{1/2}$ and $[A] = [A]_o/2$

$$\ln([A]_o/[A]_o/2) = kt_{1/2}$$

$$\ln(2) = kt_{1/2}$$

$$t_{1/2} = 0.693/k$$

The half-life does not depend on concentration for a first order reaction.

Second-Order Rate Laws

Consider the general reaction involving a reactant

$$aA \longrightarrow Products$$

the rate law becomes

$$Rate = -\Delta[A]/\Delta t = k[A]^2$$

The integrated second-order rate law has the form

$$1/[A] = kt + 1[A]_o$$

$$\text{Let } t = t_{1/2} \text{ and } [A] = [A]_o/2$$

$$1/[A]_o/2 = k_{1/2} + 1/[A]_o$$

$$2/[A]_o - 1/[A]_o = k_{1/2}$$

$$1/[A]_o = kt_{1/2}$$

$$t_{1/2} = 1/k[A]_o$$

The half-life depends on the initial concentration for a second-order reaction. Succeeding half-lives double the preceding half-life for second-order reactions. We can see this is far different than first-order reactions.

Zero-Order Rate Laws

The rate law for a zero-order reaction is

$$Rate = k[A]_o$$

$$= k$$

The rate is constant for a zero-order reaction. There is no change with concentration like first-order and second-order reactions. The integrated rate law for a zero-order reaction is

$$[A] = [A]_o - kt$$

Let $[A] = [A]_o/2$ and $t = t_{1/2}$

$$[A]_o/2 = -kt_{1/2} + [A]_o$$

$$kt_{1/2} = [A]_o/2$$

$$t_{1/2} = [A]_o/2k$$

Normally, zero-order reactions occur on metal surfaces or if an enzyme is required for a reaction to occur. An example would be a platinum surface that can accommodate a certain number of molecules.

Table 13-1: Summary of Rate Laws

Type	Zero-Order	First-Order	Second-Order
Rate Law	Rate = k	Rate = k[A]	Rate = k[A]^2
Integrated Rate Law	$[A] = [A]_o - kt$	$\ln[A] = \ln[A]_o - kt$	$1/[A] = 1[A]_o + kt$
Half-Life	$t_{1/2} = [A]_o/2k$	$t_{1/2} = 0.693/k$	$t_{1/2} = 1/k[A]_o$

Reaction Mechanisms

Chemical reactions take place by a series of steps known as the reaction mechanism. Consider the reaction

$$A + B \longrightarrow C + D$$

This reaction tells us what the reactants, products, and stoichiometry is but it does not give us information about the reaction mechanism. The mechanism

$$A + A \longrightarrow I + D \text{ Rate constant} = k_1$$

$$I + B \longrightarrow A + C \text{ Rate constant} = k_2$$

The letter I is the intermediate. It is a species that is not a reactant nor a product. It is only formed and consumed in the reaction. Each of the reactions is called an elementary step. Elementary steps are referred to as reactions where the rate law is written based on its molecularity. Molecularity is the number of species that collide to produce the reaction based on that step. Reactions in which one molecule take place is the unimolecular step. Reactions involving two and three molecules are referred to as bimolecular and termolecular.

Termolecular steps are really rare. The probability of three molecules colliding together is small. Reaction mechanisms can now be defined more precisely. It is based on a series of elementary steps that satisfies two requirements. The sum of the elementary steps gives the overall balanced equation for the reaction. The mechanism agrees with an experimentally determined rate law. Mechanism meets the second requirement by the rate-determining step. Reactions with more than one step has at least one step slower than the others. The reactants can become products only as fast as they get through the slowest step. The overall reaction cannot be faster than the rate-determining step in the sequence. Mechanisms are developed over time. The rate law is determined first. Based on knowledge and rules, the chemist comes up with possible mechanisms and by experiment eliminates those that are not correct. Mechanisms can never be proved for certainty. The best reason for a possible mechanism satisfies the two requirements given above.

Chemical Kinetic Models

The rate of chemical reactions depend on the concentration of species that are reacting with each other. The initial rate of the reaction

$$aA + bB \longrightarrow Products$$

is described by the rate law

$$Rate = k[A]^n[B]^m$$

The order of each of the reactants depend on the reaction mechanism. This is the reason reaction rates depend on concentration. It is also important to know that chemical reactions go faster when the temperature is increased. A model can be used to study the characteristics of reaction rates. The collision model is based on the idea that molecules collide to react with each other. This explains how concentration depends on reaction rates. Let us consider how this model takes into account how temperature depends on reaction rates.

The kinetic molecular theory of gases indicates if the temperature increases, molecules will have higher velocities, and there will be an increase in the frequency of collisions. Reaction rates are greater at high temperatures. There is some qualitative agreement comparing the collision model and experimental observations. The rate of a reaction is a lot smaller than the collision frequency that is calculated. Only a small fraction of collisions produce a reaction. Arrhenius came up with the idea of a threshold energy, also known as the activation energy, that has to be overcome to produce a chemical reaction. Collision model postulates state the reacting molecules have kinetic energies before the collision. The kinetic energy becomes potential energy as molecules are distorted during a collision for bonds to break and atoms to rearrange themselves into product molecules. The activated complex or transition state is found at the top of the potential energy barrier. This is where the arrangement of atoms are found. The rate only depends on the size of the activation energy, E_a.

Only collisions that have an energy greater than the activation energy can react and get over the barrier. At a low temperature, a fraction of successful collisions is small. Increasing the temperature causes the fraction of collisions that have the required activation energy to increase. These collisions increase exponentially along the temperature. The important point is that the reaction rate increases exponentially with temperature. Arrhenius came up with the idea that the number of collisions have an energy greater than or equal to the activation energy. This can be expressed as

Number of Collisions with Activation Energy = (Total Number of Collisions)$e^{-Ea/RT}$

where E_a is the activation energy, R is the universal gas constant, T is the temperature expressed in Kelvins, and the factor $e^{-Ea/RT}$ represents the fraction of collisions.

Not all molecular collisions produce chemical reactions because a minimum of energy is required for a reaction to take place. Based on experiments, observed reaction rates are a lot smaller than the rate of collisions with energy to overcome the barrier. Some collisions even though they don't have the required energy still do not produce a reaction. The answer has to do with molecular orientations for these collisions. The orientation of some of the collisions can lead to a reaction and some orientation cannot. There must be a correction factor for collisions with nonproductive molecular orientations. There are two requirements that must be satisfied for a collision to occur and rearrange to form products. The collision must have a lot of energy to produce the reaction. The collision energy has to equal or be more than the activation energy. The reactants must have a specific orientation to allow formation of new chemical bonds to form products. The rate constant can then be expressed as

$$k = zpe^{-Ea/RT}$$

The variable z is the collision frequency, p is the steric factor that is always less than one and gives us an indication of the fraction of collisions with effective orientations, and $e^{-Ea/RT}$ is the fraction of collisions that have enough energy to make a reaction. The rate constant is normally written in the form

$$k = Ae^{-Ea/RT}$$

This is called the Arrhenius equation. The constant A represents the frequency factor. Note that

$$A = zp$$

If we take the natural logarithm of both sides of the Arrhenius equation we have

$$\ln(k) = -E_a/R(1/T) + \ln(A)$$

Since most rate constants obey the Arrhenius equation to a close approximation, it tells us that the collision model is reasonable to use. A common procedure for determining the activation energy involves measuring the rate constant, k, at many temperatures and then plotting ln(k) versus 1/T. Another way to calculate activation energy is knowing the values of k at only two temperatures based on the following formula.

$$\ln(k_1) = -E_a/RT_1 + \ln(A)$$

$$\ln(k_2) = -E_a/RT_2 + \ln(A)$$

$$\ln(k_2) - \ln(k_1) = [-E_a/RT_2 + \ln(A)] - [-E_a/RT_1 + \ln(A)]$$

$$\ln(k_2) - \ln(k_1) = -E_a/RT_2 + E_a/RT_1$$

$$\ln(k_2/k_1) = E_a/R(1/T_1 - 1/T_2)$$

Catalysis

Reaction rates increase pretty good with temperature. Sometimes, if a reaction does not happen fast enough at normal or room temperature, then the reaction rate can speed up by increasing the temperature. Sometimes, this will not always work. The body contains substances called enzymes which increase the reaction rates. Almost all biologic reactions get help from enzymes. Sometimes increasing the temperature of a reaction is not the best way to go because it might cost too much money and too much energy. Reactions can proceed rapidly with the use of a catalyst at a low temperature and bring down the production costs. A catalyst is a substance that makes a reaction proceed faster without itself being consumed. Most industrial processes make use of a catalyst. Let us take a closer look at how a catalyst works. An energy barrier has to be overcome for a reaction to proceed. Finding a new pathway that has a lower activation energy can make a reaction proceed faster. Catalysts allow reactions to occur by having a lower activation energy. A larger portion of collisions take place, and the reaction rate is increased. A catalyst only lowers the energy of activation for a reaction. It does not affect the difference in energy between reactants and products.

Heterogeneous catalysis takes place when gaseous reactants are adsorbed on the surface of a solid catalyst. One substance being collected on the surface of another substance is referred to as adsorption. Adsorption can also be thought of as one substance penetrating into another substance. An example would be water being absorbed by a sponge. If we consider the reaction of ethylene with hydrogen gas to form ethane, this process can be considered as a heterogeneous catalysis. This can take place in four different steps. The first step involves adsorption and activation of the reactants, adsorbed reactants migrating on the surface, reaction of adsorbed substances taking place, and desorption of the products. A homogeneous catalyst takes place in the same phase as the molecules react with each other. An example would be nitric acid catalyzing ozone production in the troposphere, but also catalyzing the decomposition of ozone in the upper atmosphere.

Conclusion

Chemical reactions might have a tendency to occur, but that does not mean the reaction will occur fast. Chemical kinetics is about factors that control the rate of chemical reactions. The rate of a reaction is considered to be the change in concentration of reactant or product per unit of time. Kinetic conditions take place where the reverse reaction is not significant. Only the forward reaction is important for finding the rate. Two kinds of rate laws are the differential rate law and the integrated rate law. Mechanisms are thought of a series of elementary steps for a reaction. For each of these steps, the rate law is written based on molecularity. This means the number of species that collide in order to make a reaction to occur. An unimolecular step is considered to be first order, and a bimolecular step is considered to be second order. If we accept a mechanism, the sum of the elementary steps gives the overall balanced equation. The mechanism should agree with the rate law for the experimentally determined rate law. Multiple step reactions have one step slower than the others, and this is called the rate-determining step. Rate laws can be summarized as zero-order, first-order, and second-order.

Chapter 14

Equilibrium

Laboratory Balance and Glassware

When we were doing stoichiometry calculations, we have made the assumptions that reactions proceed to completion. In other words, one of the reactants runs out. A lot of reactions proceed to completion but not all of them. This is the case where we feel the limiting reactant is negligible. However, some of the reactions do not go to completion. Sometimes, a system reaches chemical equilibrium. This is the state where the concentrations of all reactants and products stay constant. Chemical reactions carried in a closed vessel reaches equilibrium. If the equilibrium position favors the products, the reaction appears to have gone to completion. The equilibrium position therefore lies far to the right. However, some reactions such as solids decomposing slightly to products usually has the equilibrium being towards the left. Let us look at chemical systems and the processes of equilibrium. We will soon learn how to calculate reactants and products based on equilibrium expressions.

Conditions of Equilibrium

There are no changes in the concentration of reactants or products when the reaction system is in equilibrium. It is important to know the reaction did not stop. The equilibrium process is dynamic. It can be represented as the following equation.

$$A + B \rightleftharpoons C + D$$

The concentration of reactants and products stay constant during the equilibrium process. The forward and reverse reaction rates become equal. Molecules react by colliding with one another. The more collisions there are the faster the reaction. Reaction rates depend on concentration. Sooner or later, the concentrations reach a certain level where the forward and reverse rates become equal. We say the system is in a state of equilibrium. Equilibrium depends on initial concentrations, energies of reactants and products, and the organization of reactants and products.

The concentration of reactants and products stay the same when they are mixed together because the system is at chemical equilibrium and the forward and reverse reactions are slow such that the system moves toward equilibrium at a rate that we cannot detect.

The Equilibrium Constant

The law of mass action is a general description of the equilibrium condition. Consider the following reaction

$$aA + bB \rightleftharpoons cC + dD$$

where A, B, C, and D are chemical species and a, b, c, and are coefficients in the balanced equation. The law of mass action is represented as follows

$$K = [C]^c[D]^d/[A]^a[B]^b$$

Square brackets indicate concentrations for each of the chemical species that is at equilibrium. The variable, K, is considered to be the equilibrium constant. If we know the equilibrium concentration for the reaction components, then we can calculate the value of K. Equilibrium constants written without units has to do with corrections for the non ideal behavior of substances taking part in the reaction.

$$K' = 1/K$$

Multiplying a factor x in the original equation gives us

$$K'' = K^{\Delta x}$$

The law of mass action describes the behavior of equilibrium in solution and in the gaseous state. The equilibrium constant always has the same value for a given temperature. It does not matter the amount of gases mixed together to begin with. Although the ratio of the equilibrium expression is constant, the equilibrium concentrations are not all the same. A set of equilibrium concentrations is called an equilibrium position. There is only one equilibrium constant at a specific temperature, but there are a lot of equilibrium positions. The equilibrium position depends on initial concentrations. However, the equilibrium constant does not depend on the initial concentrations.

Equilibrium Expressions that involve Pressure

The relationship between pressure and concentration of gas is

$$PV = nRT$$

$$P = (n/V)RT$$

$$P = CRT$$

The variable C is the molar concentration of the gas. The value, K_c, would be the equilibrium expression for concentration. The value, K_p, represents the equilibrium constant in terms of partial pressures. The following equation can be related to each other based on the following general equilibrium condition.

$$aA + bB \rightleftharpoons cC + dD$$

$$K_p = K(RT)^{\Delta n}$$

where $\Delta n = (c + d) - (a + b)$

Heterogeneous Equilibria

Heterogeneous equilibria is equilibria that involve more than one phase. If we were to look at the chemical equilibrium for the following reaction

$$CaCO_3(s) \rightleftharpoons CaO(s) + CO_2(g)$$

we find experimentally the position of the heterogeneous equilibrium does not depend on the amount of solids or liquids present. The concentration of pure solids and pure liquids do not change. The concentration of pure solids or pure liquids are not included in an equilibrium expression. The concentration of solutions and gases vary on the other hand. We can represent the equilibrium for carbon dioxide as

$$K = [CO_2]$$

Extent for a Reaction

The tendency for a reaction to occur has to do with the size and extent of the equilibrium constant. The equilibrium value greater than 1 means the reaction will be mostly products. In other words, the equilibrium lies toward the right hand side of the equation. Reactions that have large equilibrium constants go to completion. Small values of K mean the equilibrium position lies to the left where the reactants are at. The reaction barely occurs. The size of K and the time that is required to go to equilibrium has nothing to do with each other. The time needed for equilibrium depends on the rate of the reaction. This depends on the size of activation energy. The size of K depends on the difference in energy between products and reactants.

When reactants and products mix, we like to know whether the mixture stays at equilibrium or if the direction of equilibrium is needed for it to reach equilibrium. If the concentration of either the reactant or product is zero, the system shifts in the direction towards the missing component. On the other hand, if the initial concentrations are not zero, it is not easy to determine which side the equilibrium will move. If we want to determine the direction of the shift, we look at the reaction quotient which is symbolized as Q. We get the reaction quotient by using the Law of Mass Action. Initial concentrations are used in place of equilibrium concentrations. If we want to know the direction the equilibrium will shift, then we have to look at Q and K values. There are three possibilities to look at. The first case is when the system is at equilibrium where there is no shift. This is the case when Q equals K. The second case is when the initial concentration of products to initial concentration of reactants would be large if Q is greater than K. In order for the system to reach equilibrium, the system will have to shift to the left going from products to reactants. The third case is when the initial concentration of products to initial concentration of reactants is small. The system will go to the right. Reactants end up forming products for the system to reach equilibrium.

The tendency for a reaction to occur depends on the magnitude of the equilibrium constant. If the K value is greater than 1, the equilibrium will shift to the right where the products are at. However, if the K value is smaller than 1, the equilibrium will shift to the left where the reactants are at. The magnitude of K and its time are not directly related to each other. Time for equilibrium to occur depends on the rate of the reaction. The reaction rate depends on the energy of activation, E_a. The K value is determined based on the difference in energy between products and reactants, ΔE. We need to know the direction of equilibrium or if the mixture is at equilibrium when reactants are mixed together. The direction of equilibrium will shift towards the side with the smaller concentration. The reaction quotient, Q, determines the direction of the shift. We obtain the reaction quotient by applying the Law of Mass Action. Initial concentrations instead of equilibrium concentrations are used for the reaction quotient. Comparing the values of Q and K will tell us the direction the equilibrium will shift. If Q = K, then the system is at equilibrium and there will be no shift. If Q is greater than K, then the system will shift to the left until an equilibrium condition is achieved. If Q is less than K, the system will shift to the right until an equilibrium condition is met.

Le Chatelier's Principle

Le Chatelier's principle states that if a system is disturbed during equilibrium, then the system will adjust itself by shifting in the direction of equilibrium back to its normal state. Another way stating Le Chatelier's principle is if a reactant or product is added to the reaction system during equilibrium, then the equilibrium shifts towards the lower concentration of the component. The opposite occurs if the component is removed. Adding inert gases increase the total pressure on the system. But there is no effect on the concentrations or partial pressures of reactants or products. If the volume of container that holds a gaseous system gets reduced, the system itself will reduce its own volume. The total number of gaseous molecules will be decreased. According to the ideal gas equation

$$V = (RT/P)n$$

This is when T and P are constant and $V \propto n$. If we have the general reaction

$$A + B \rightleftharpoons C$$

and decrease the volume in the container with both A and B, then C molecules will get produced and the direction of equilibrium will shift to the right. If the container's volume is increased, then the system will shift to increase its volume. Increasing the volume of the system of C will shift the equilibrium to the left. This will increase both A and B.

Effects of Temperature Change

The conditions just described changes the equilibrium position. The equilibrium constant does not change. However, the value of K changes with temperature. Suppose we have the general reaction

$$A \rightleftharpoons B + Energy$$

The direction of shift will be towards the side that consumes energy. In this case, it will go to the left. This ends up decreasing the concentration of B, increasing the concentration of A, and therefore decreasing the value of K. When temperature increases, the value of K decreases. If we have the reaction

$$Energy + A \rightleftharpoons B$$

Increasing temperature causes the equilibrium to shift towards the right. The value of K increases. It is worth mentioning again that Le Chatelier's principle does not predict the size of K nor its change. It predicts only the direction of the change.

Effects of Concentration Change

If we consider the reaction

$$3H_2(g) + N_2(g) \longrightarrow 2NH_3(g)$$

and if we decide to add more N_2 to the equilibrium mixture, the reaction will shift to the right. More N_2 and H_2 molecules collide with each other. The forward reaction rate increases, and the rate of formation of ammonia molecules increase. An excess of ammonia molecules will make the reaction go back in reverse. The forward and reverse reactions will eventually become equal. The system will be at its new equilibrium position. Le Chatelier's principle can also be stated if one of the reactants or products is added to the reaction system at equilibrium. The equilibrium position shifts to lower the concentration of the component. Removing a component makes the opposite effect to occur.

Effects of Pressure Changes

There are three ways to change pressure that involve gas components. A gas reactant or product can be added or removed, an inert gas can be added, or the volume of the container can be changed. Adding an inert gas does not affect the equilibrium position. Adding an inert gas increases the total pressure, but it does not affect the concentrations or partial pressures of reactants or products. The system stays at its original equilibrium position. Changing the volume of the container changes the concentrations or partial pressures of reactants and products. The value, Q, can be calculated, and the direction of the shift can be calculated. If the volume of the container that holds a gaseous system gets reduced, then the system reduces its own volume. Eventually, the total number of gaseous molecules decrease. The ideal gas law can be rearranged to give volume at a constant temperature and pressure. Volume is proportional to the number of moles.

If we consider the reaction for the synthesis of ammonia in a closed container and reduce its volume, more of the reactant molecules of hydrogen gas and nitrogen gas produce ammonia molecules. The total number of gaseous molecules get reduced. The equilibrium will be farther to the right where the smaller number of gaseous molecules are present in the balanced equation. However, if you increase the volume of the container, there will be a shift towards increasing its volume. Increasing the volume for the synthesis of ammonia will cause a shift to the left to increase the gaseous molecules present.

Conclusion

When doing stoichiometry problems, reactions are assumed to run to completion. If a reaction is carried out in a closed vessel, the system is at chemical equilibrium. This is where reactants and products stay constant over time. Equilibrium concentrations can be small, but it can never reach zero. Equilibrium is dynamic where reactants are constantly forming products, and vice versa. Equilibrium have rates becoming the same for forward and reverse reactions. The law of mass action describes equilibrium. Consider the general reaction of the type

$$aA + bB \rightleftharpoons cC + dD$$

the equilibrium expression gives

$$K = [C]^c[D]^d/[A]^a[B]^b$$

The variable, K, is the equilibrium constant. For reactions at a specific temperature, there is a value for the equilibrium constant, but there are many possible equilibrium positions. Equilibrium positions have concentrations that satisfy the equilibrium expression. Equilibrium position depend on initial concentrations. Equilibrium position do not depend on the amount of pure solid or pure liquid. These substances are not included in the equilibrium expression. Gas-phase reactions have equilibrium positions that are described by partial pressures. Equilibrium for pressure can be calculated based on the equation $K_p = K(RT)^{\Delta n}$ where Δn is the sum of the coefficients of products minus reactants for gases.

Homogeneous equilibria have to do with reactants and products being in the same phase. Other systems involve different phases, and their equilibria is heterogeneous. The K value predicts whether a reaction can occur. A small K value means the equilibrium position is towards the left. The reaction does not occur to any significant extent. A larger K value means the reaction more likely goes to completion. The equilibrium position is towards the right hand side. The K value, however, does not contain information if the equilibrium will be fast or slow. The reaction quotient Q uses the Law of Mass Action for initial concentrations. The Q and K values are used to compare the direction a chemical system will shift to reach equilibrium. If Q and K are equal, then the system is at equilibrium. If Q is less than K, then the system shifts to the right to reach equilibrium. If Q is greater than K, then the system shifts to the left to reach equilibrium.

In order to find concentrations for a given equilibrium position, it is a good idea to start with the given initial concentrations or pressures, write the change to reach equilibrium, and apply the change to initial concentrations or pressures for the equilibrium expression. Le Chatelier's principle predicts the effects of changes in concentration, pressure, and temperature for equilibrium. If we disturb the system, then the system will adjust to restore the equilibrium in the proper direction.

Chapter 15

Acids, Bases, and Buffers

Beautiful Bicolor Pink and Purple Mophead Hydrangea Flower Head

Acids and bases have been common since ancient times. They are applied to chemical equilibria involving proton-transfer reactions. Acid-base chemistry is used for everyday applications. Biological processes carefully control the acidity of our blood. Small changes in acidity can cause serious illness and possibly death. Other life forces too need to control or regulate their acidity in order to survive. Acids and bases are also used in industry. Sulfuric acid is one of the top leading chemicals manufactured in the United States of America. We need sulfuric acid to produce fertilizers, steel, and other materials such as polymers. Acids can also be a problem. Acid rain, for example, has caused destructive damage to buildings and polluted our environments. Loss of habitat life has occurred due to acid rain. Long ago during the early days, acids were recognized as being sour. Vinegar is a dilute solution of acetic acid. It tastes sour. Citric acid tastes sour too due to the taste of a lemon. These are food items we taste almost on a daily basis. However, do not taste chemicals. They are poisonous. Bases, on the other hand, have a bitter taste and feel slippery.

Svante Arrhenius recognized the importance of acids and bases. He did experiments on electrolytes. He came up with the idea that acids produce hydrogen ions in aqueous solution. He also came up with the idea that bases produce hydroxide ions. This lead to the definition of Arrhenius acids and Arrhenius bases. But this definition is narrow. It applies only to aqueous solutions and restricts itself only to one kind of base; that is, the hydroxide ion. Johannes Bronsted came up with a more general definition of acids and bases. Thomas-Lowry also contributed his expertise to this general definition. The Bronsted-Lowry Model states that an acid is a proton donor, and a base is a proton acceptor.

Water can act both as an acid and as a base. It is considered to be amphoteric. Water can act as a base since oxygen on the water molecule has two lone pairs. The lone pairs can form a covalent bond with a hydronium ion. When water reacts with hydrobromic acid, the products are the hydronium cation and the bromide anion. We represent the abbreviation as "aq" to mean the substance is hydrated. Water is also a polar molecule that pulls off a proton from an acid such as hydrobromic acid. A conjugate base is defined as the entire molecule or compound with the proton lost. A conjugate acid is formed when a proton is transferred to a base. Conjugate acid-base pairs are two substances that are related to each other when a proton is donated or accepted. We can represent the general reaction of an equation to be

$$HA(aq) + H_2O(l) \rightleftharpoons H_3O^+ + A^-(aq)$$

The conjugate acid-base pairs are HA and A^- as well as H_2O and H_3O^+. Based on this reaction, the proton competes between the two bases H_2O and A^-. If water is a stronger base than the anion, it will pull off a proton from HA and the equilibrium position will be far to the right. The dissolved acid will be in the ionized form. However, if the anion is a stronger base than water, the equilibrium will be far to the left. Most of the acid will remain as the unionized form. We can write an equilibrium expression for the reaction written above

$$K_a = [H^+][A^-]/[HA]$$

The value, K_a, is called the acid dissociation constant. Remember that the hydronium cation can also be represented as H^+ or H_3O^+. The two notations represent the hydrated proton.

Recall that the concentration of a pure solid or a pure liquid is omitted from the equilibrium expression. If the solution is dilute, the concentration of liquid water remains the same. This is the reason why we don't include the concentration of water in the equation. The equilibrium expression can be rewritten as

$$HA \rightleftharpoons H^+(aq) + A^-(aq)$$

The value, K_a, is also called the equilibrium constant. A proton gets removed from HA to form the conjugate base, A^-. The value, K_a, is used for this kind of reaction. Reactions can also take place in the gas phase. If ammonia and hydrochloric acid both in the gas phase react, they can form solid ammonium chloride.

Strength of Acids

We can define the strength of an acid by its equilibrium position of a dissociation reaction. We have seen the general form of this kind of reaction. Strong acids have equilibrium positions shifting far to the right. Almost all of the HA molecules becomes dissociated or ionized at equilibrium. This is a relationship of acid strength and its conjugate base. Strong acids give weak conjugate bases. Remember a weak conjugate base has a low affinity for a proton. Strong acids have conjugate bases that are a lot weaker than water. Therefore, water molecules attract hydronium ions more easily.

Weak acids are acids for which the equilibrium's position is toward the left. In solution, most of the acid is present as HA at equilibrium. Weak acids dissociate to a small extent as HA at equilibrium. Weak acids have conjugate bases that are a lot stronger than water. Water molecules have a difficult time removing a hydronium cation from a conjugate base. Weaker acids mean the conjugate base is a lot stronger. Sulfuric acid, hydrochloric acid, nitric acid, and perchloric acid are considered to be strong acids. Sulfuric acid is considered to be a diuretic acid which means it is an acid with two acidic protons. Sulfuric acid is a strong acid which means it can dissociate 100% in water. Consider the reaction of sulfuric acid dissociating to the hydronium cation and the hydrogen sulfate anion. The anion hydrogen sulfate is considered to be a weak acid that is in equilibrium with another hydronium ion and the sulfate anion.

Oxyacids have the acidic proton attached to an oxygen atom. The weak acids phosphoric acid, hypochlorous acid, and nitrous acid are also considered to be oxyacids. Organic acids have a carbon atom backbone that contains the carboxyl group (-COOH). Carboxyl group acids are weak. Acetic acid and benzoic acid are also considered to be weak acids. The other hydrogens not bonded to the carboxyl group are considered to be not acidic. They don't form hydronium ions in water. Hydrohalic acids are represented as HX where X is a hydrogen atom. Examples include hydrochloric acid, hydrobromic acid, and hydroiodic acid. The three hydrohalic acids are considered to be a weak acid. Values of K_a for strong hydrohalic acids cannot be measured accurately since they completely dissociate in water.

Water being Amphoteric

Amphoteric substances act either as acids or bases. An example of an amphoteric substance is water. The auto ionization of water involves transferring a proton from one of the water molecules to another water molecule to give a hydronium ion and a hydroxide ion. We can represent the following reaction as

$$H_2O + H_2O \rightleftharpoons H_3O^+ + OH^-$$

One of the water molecules act as an acid ready to give a proton, whereas another water molecule acts as a base ready to accept a proton. Autoionization can also occur for two ammonia molecules producing one of the products as an ammonium cation and another product as an amide anion. An equilibrium expression for the auto ionization of water is

$$K_w = [H^+][OH^-]$$

This is when K_w is considered to be the ion-product constant or the dissociation constant for water. At 25°C, the concentrations of hydronium and hydroxide ions are equal.

$$[H^+] = [OH^-] = 1.0 \times 10^{-7} \text{ M}$$

Plugging the concentrations of each of them for Kw gives

$$K_w = [H^+][OH^-]$$

$$K_w = (1.0 \times 10^{-7})(1.0 \times 10^{-7})$$

$$K_w = 1.0 \times 10^{-14}$$

Neutral solutions consist of $[H^+] = [OH^-]$, acidic solutions consist of $[H^+] > [OH^-]$, and basic solutions consist of $[OH^-] > [H^+]$.

pH of Weak Acid Solutions

Let us take a look at the weak acid hydrofluoric acid. As always, it is important to write the major species in solution. We know HF is a weak acid because the K_a value is small. The value is about 7.2 x 10^{-4}. The molecule will only be dissociated slightly, and therefore the reaction will be at equilibrium. In other words, the molecule HF will more likely be intact than separated into ions in solution. Although HF is considered to be a weak acid, it is a lot stronger than water. Eventually towards the end of our calculations we can get an approximate value for the concentrate of hydrofluoric acid. We can use this value to calculate the percent dissociation for a weak acid. Percent dissociation is the amount dissociated in moles per liter divided by its initial concentration in moles per liter multiplied by 100%. The equation can be written as

Percent Dissociation = (amount dissociated/initial concentration) x 100%

The percent dissociation for weak acids increase when the acid becomes more dilute. For a weak acid HA, the concentration of the hydronium ion decreases when the initial concentration of the original acid decreases. The percent dissociation increases when the initial concentration of the original acid decreases.

pH of Strong Acid Solutions

It is important to pay attention to the components in solution when dealing with acid-base equilibria. Let us take a look at the bottle 1.0 M HCl. Although the bottle states 1.0 M HCl, there are no HCl molecules present in the solution. There are only H^+ and Cl^- ions in solution. These are the major species present in solution. The other major species present in solution is water. The solution is considered to be very acidic. Only very tiny amounts of hydroxide ions are present in solution. It is considered to be a minor component in solution. When working with acid-base problems, it is important to write the major species in solution. Other strong acids include nitric acid, sulfuric acid, and perchloric acid. We are interested calculating the pH so we will focus on the major species hydronium in solution. The water molecule also furnishes hydronium by the process of auto ionization. The dissociation reaction is written as

$$H_2O(l) \rightleftharpoons H^+(aq) + OH^-(aq)$$

Adding 1.0 M HCl to the solution causes the equilibrium to shift to the left since more hydronium ions are added to the right of the equation. Water contributes a negligible amount of hydronium ions compared to the 1.0 M hydronium ions from HCl dissociating in solution. The concentration of the hydronium ion is still 1.0 M. The pH is therefore calculated as

$$pH = -log[H^+]$$

$$pH = -log(1.0)$$

$$pH = 0$$

The pH Scale

The concentration of the hydronium cation in aqueous solutions is very small. The pH scale provides a useful way to represent the acidity of a solution. The pH is considered to be a log scale that is based on 10. The equation has the form

$$pH = -\log[H^+]$$

Like any calculations, we have to watch out for significant figures for logarithms. The number of decimal places in a log equals the number of significant figures in the original number. Log scales can also be used to represent other quantities such as

$$pOH = -\log[OH^-]$$

or

$$pK_a = -\log K_a$$

The pH is a scale based on 10. The pH changes by 1 for every power of 10 change in the hydronium ion concentration. As pH increases, the concentration of the hydronium ion decreases. Similarly, if pH decreases, the concentration of the hydronium ion increases. A solution's pH can be measured with a pH meter. A pH meter is an electronic device that has a probe. It can be inserted into a solution to determine its unknown pH. There is an acidic aqueous solution in the glass membrane for the hydronium ions to migrate. If the pH of the unknown solution is different from the pH of the solution in the probe, an electrical potential takes place on the meter. The log form of the expression for the ion-water constant can be rearranged algebraically into the following forms

$$K_w = [H^+][OH^-]$$

$$\log K_w = \log[H^+] + \log[OH^-]$$

$$-\log K_w = -\log[H^+] - \log[OH^-]$$

$$pK_w = pH + pOH$$

$$\text{At } 25°C, K_w = 1.0 \times 10^{-14}$$

$$pK_w = -\log(1.0 \times 10^{-14})$$

$$pK_w = 14.00$$

Therefore, the following equation can be written as

$$pH + pOH = 14.00$$

Bases

An Arrhenius base is considered to be a substance that produces hydroxide ions in aqueous solution. The Bronsted-Lowry Model of a base is considered to be a proton acceptor. Examples of both an Arrhenius base and a Bronsted-Lowry base are solid hydroxide and potassium hydroxide. In the solid state, hydroxide ions are locked into the solid lattice. When sodium hydroxide or potassium hydroxide are mixed in water, they dissociate completely into its cations and anions. Group 1A and Group 2A hydroxides are considered to be strong bases. Alkaline Earth hydroxides are not too much soluble. Usually, they are used when solubility is not important. Examples of antacids include aluminum hydroxide and magnesium hydroxide. Slaked lime is considered to be calcium hydroxide. It is used to scrub gases and remove sulfur dioxide from exhausts of power plants and factories. Slaked lime is also used in water treatment plants in order to soften hard water. The lime-soda process is one of the softening processes in water treatment plants. Lime and soda ash are both added to water. Lime reacts with water to form slaked lime which then reacts with soda ash and calcium ions in hard water to make calcium carbonate.

It is important to note that a base does not need to have a hydroxide ion. However, substances such as these when dissolved in water increase the hydroxide ion concentration in water. Ammonia reacts with water to produce the ammonium cation and hydroxide anion. The ammonia molecule acts as a base and accepts a proton. Water acts as an acid. Ammonia does not contain a hydroxide ion concentration, but it produces the hydroxide ion by reacting with water in an aqueous solution. Let us take a look at how ammonia acts as a base. Ammonia has one unshared lone pair of electrons that removes one of the hydrogen atoms to form the ammonium cation and the hydroxide anion. The general reaction between a base and water is

$$B(aq) + H_2O(l) \rightleftharpoons BH^+(aq) + OH^-(aq)$$

| Base | Acid | Conjugate Acid | Conjugate Base |

The equilibrium constant for this reaction is

$$K_b = [BH^+][OH^-]/[B]$$

The value, K_b, refers to the reaction of a base reacting with water to form the hydroxide anion and the conjugate acid. Bases such as B compete with the hydroxide ion concentration. The K_b values are small and are considered to be weak bases.

Polyprotic Acids

Polyprotic acids can give more than one kind of proton. Examples of polyprotic acids are sulfuric acid and phosphoric acid. Polyprotic acids dissociate in a stepwise manner. One proton is removed one at a time. An example of a diprotic acid is carbonic acid. An example of a diprotic acid is sulfuric acid. Sulfuric acid is a strong acid for the first dissociation step, but it is weaker in the second dissociation step. If the second step makes a big contribution to [H⁺], then that means we have a dilute solution of sulfuric acid. The quadratic equation has to be used to obtain the H⁺ concentration. An example of a triprotic acid is phosphoric acid. For weak polyprotic acids, the value of K_a decreases as each hydrogen cation is removed from phosphoric acid. The acid gets weaker at each successive step. The second and third proton gets lost less readily than the first proton. This makes sense since the negative charge on the acid increases. It gets to be more difficult to remove the positively charged proton. Normally a polyprotic acid's first dissociation step is important when determining the pH.

Properties of Salt

Salt is made up of an ionic compound. Adding salt to water causes the ions to break apart into its cation and anion. They move around independently in dilute solutions. The ions can act as acids or bases. The conjugate base of a strong acid does not have an affinity for protons in water. This explains the reason why strong acids dissociate completely in aqueous solution. Anions such as Br⁻ or I⁻ in an aqueous solution do not combine with H⁺. These ions do not affect the pH. Cations such as Li⁺ and Na⁺ do not have an affinity for H⁺. These ions also cannot produce H⁺. They do not affect the pH of an aqueous solution. We can conclude that salts which consist of cations of strong bases and anions of strong acids do not affect the pH when dissolved in water. Aqueous solutions of salts like NaCl or KNO_3 are considered to be neutral. They have a pH of 7. A strong acid and a strong base give a salt in a neutral solution.

An aqueous solution of sodium acetate consist of the major species Na⁺, $C_2H_3O_2^-$, and H_2O. The sodium cation is not an acid or a base. The acetate anion is the conjugate base of acetic acid which is a weak acid. The acetate anion acts as a base and has a strong affinity for a proton. Water acts as a weak amphoteric substance. The pH of the solution is determined by the acetate anion. The acetate anion is a base and reacts with a proton donor. The reaction where an acetate anion reacts with water gives acetic acid and a hydroxide anion. The reaction gives a basic solution. A base reacts with water to give the hydroxide anion and a conjugate acid. A basic equilibrium constant can be obtained. If you multiply the acidic equilibrium constant and the basic equilibrium constant you will get the ion-product constant for water.

For a salt whose cation is neither acidic nor basic and whose anion is a conjugate base of a weak acid, the aqueous solution is considered to be basic. A solution is basic if the anion of a salt is the conjugate base of a weak acid. Let us look at the strength of bases in solution. Aqueous hydrocyanic acid can react with liquid water to form an equilibrium with the hydronium cation and the cyanide anion. Hydrocyanic acid is considered to be a weak acid. The cyanide anion is considered to be a strong base. It has a high affinity for the hydronium cation when compared to water. Both substances are competing in solution.

The cyanide anion also reacts with water in solution. The cyanide anion is considered to be a weak base. In this reaction, the cyanide anion competes with the hydroxide anion for the hydronium cation instead of competing for water molecules. The hydroxide anion is a stronger base than the cyanide anion, and the cyanide anion is stronger base than water. Salts can also produce acidic solutions in dissolved water. Solid ammonium chloride dissolved in water gives the ammonium cation and the chloride anions. The ammonium cation behaves as a weak acid. The chloride anion does not have an affinity for the hydronium cations in water. It has no effect on the pH of a solution. Salts that contain anions which are not bases and cations whose conjugate acids of a weak base give acidic solutions.

Another kind of salt that gives an acidic solution is a highly charged metal cation. If we consider solid aluminum chloride dissolved in water, the following solution is acidic. The Al^{3+} is not a Bronsted-Lowry acid. However, the hydrated cation $Al(H_2O)_6^{3+}$ is considered to be a weak acid. The metal cation has a high charge that polarizes O-H bonds in the water molecules. Higher charges on metal cations cause the stronger acidity of the hydrated ion. Let us also look at the compound ammonium acetate. Both ions affect the pH of an aqueous solution. If the K_a value for the acidic ion is bigger than the K_b value for the basic ion, the solution is considered to be acidic. If the K_b value for the basic ion is bigger than the K_a value for the acidic ion, the solution is considered to be basic. A neutral solution has equal K_a and K_b values.

Structural Effects

A substance dissolved in water can produce an acidic solution if it can donate protons or it can produce a basic solution if it can accept protons. Let us take a look at the structural properties to see why it behaves as an acid or as a base. In general, a molecule that contains a hydrogen atom can behave as an acid. However, there are molecules that show incredibly weak acidic properties such as the carbon bonded to hydrogen. These do not produce acidic aqueous solutions because the carbon bonded to hydrogen is strong and nonpolar. It does not have the tendency to donate protons. Even though a molecule such as HBr is stronger than the CH bond, it is more polar and the molecule more readily dissociates in water. The strength and polarity of the bond determines whether the molecule behaves as a Bronsted-Lowry acid.

The polarity of the hydrogen halides decrease from hydrogen fluoride to hydrogen chloride to hydrogen bromide to hydrogen iodide. The molecule hydrogen fluoride is the most polar, whereas hydrogen iodide is the least polar. Electronegativity decreases going down a group. The molecule hydrogen fluoride is a weak acid even though the bond is highly polar. The rest of the hydrogen halides such as hydrogen chloride, hydrogen bromide, and hydrogen iodide are strong acids.

Oxyacids are another group of acids. The strength of an acid increases with the number of oxygen atoms attached to the central atom. Oxygen is electronegative and is able to withdraw electrons away from the bond. A proton is more readily to be pulled off by a molecule with increasing number of oxygen atoms attached. This process is similar to hydrated metal ions. Highly charged metal cations can produce acidic solutions. Examples of highly charged metal ions are Fe^{3+} and Al^{3+}. Water molecules can act as acids. The electronegative oxygen atoms of water molecules orients itself around the highly charged metal cation. There is an increase in attraction of electrons to the positive metal cation. A higher charge on the metal cation gives a more acidic hydrated ion. If we take a look at the general structure of an oxyacid H-O-X, the better the ability of X to draw electrons to itself, the more acidic the molecule. Acid strength depends on the electronegativity of X. There is a strong correlation between the electronegativity of X and acid strength for oxyacids.

Oxides

Molecules that contain H-O-X behave as acids, and the strength of an acid depends on the electron-withdrawing ability of X. However, substances containing the H-O-X grouping can also behave as bases if the hydroxide anion is produced. The nature of the O-X bond determines the behavior whether the solution will become acidic or basic. If X is highly electronegative, the O-X bond will be both strong and covalent. When it is dissolved in water, the O-X bond stays intact. The O-H bond becomes weak and releases a proton. If X has a low electronegativity, the O-X bond is considered to be ionic. It is more likely to be broken in water. Sodium hydroxide and lithium hydroxide dissolve in water and give the metal cation and a hydroxide anion. These principles can be used to explain acid-base behavior of oxides when dissolved in water. Sulfur trioxide dissolved in water produces sulfuric acid. The O-S bonds remain attached, but the weak O-H bonds break apart and give protons. Another example includes carbon dioxide combined with water to make carbonic acid. They are considered to be covalent oxides that dissolve in water to give acidic solutions. They are also considered to be acidic oxides. When an ionic oxide dissolves in water, a basic solution is produced. An example would be to combine calcium oxide with water to produce calcium hydroxide. Basic oxides are normally produced from the cations of Groups 1A and 2A.

Lewis Acid-Base Model

G.N. Lewis during the early 1920s suggested a model for acid-base behavior. A Lewis acid is considered to be an electron-pair acceptor, whereas a Lewis base is considered to be an electron-pair donor. Lewis acids have empty atomic orbitals that is used to accept an electron pair from a molecule that has lone pair electrons. Lewis bases are molecules that have lone pair of electrons. An example would be to combine the hydronium cation and ammonia to form the ammonium cation. The Lewis acid is the hydronium cation and the Lewis base is the ammonia molecule. Another example is the reaction between boron trifluoride which is electron-deficient, and it completes its octet by reacting with ammonia that has a lone pair of electrons. Boron trifluoride is reactive since it has an empty p orbital. It is therefore a strong Lewis acid.

<u>Conclusion</u>

Arrhenius acids produce hydronium cations in solution, whereas Arrhenius bases produce hydroxide anions in solution. A Bronsted-Lowry acid is a proton donor, whereas a Bronsted-Lowry base is a proton acceptor. Water molecules act as Bronsted-Lowry bases when it accepts protons from acids to form hydronium cations. Conjugate bases is the rest of the molecule after a proton is lost. Conjugate acids form when protons are transferred to the base. These two substances considered together are called conjugate acid-base pairs. The equilibrium expression for acids dissociating in water can then be calculated. The symbol H^+ is the same as H_3O^+. The concentration of water is not included because it is assumed to be constant. The value K_a is called the acid dissociation constant. Acid strength depends on the position of the dissociation equilibrium. Small values of K_a denote weak acids. High values of K_a denote strong acids. They lie far to the right of the equilibrium. Strong acids have weak conjugate bases. Weak conjugate bases have low affinity for protons. Acid strength is inversely related to conjugate base strength.

Water molecules are considered to be amphoteric. They can behave either as acids or as bases. The auto ionization of water shows this. A water molecule transfers a proton to another water molecule to make a hydronium cation and a hydroxide anion. The ion-product for water can then be calculated. At room temperature, the ion-product for water is 1.0×10^{-14}. Pure water therefore have equal concentrations of hydronium and hydroxide ions which are each 1.0×10^{-7}. Acidic solutions have hydronium cations concentrations greater than hydroxide anion concentrations. Basic solutions have hydroxide anion concentrations greater than hydronium cation concentrations. Neutral solutions have equal concentrations of hydronium and hydroxides.

The pH scale describes the hydronium cation concentrations in aqueous solutions. The equation can be represented as $pH = -\log[H^+]$. The value of pH is considered to be a log scale based on 10. The pH changes by 1 for each change of power by 10 in the hydronium cation concentrations. The pH value decreases as the hydronium ion concentration increases. Likewise, the pH value increases as the hydronium cation concentration decreases. Percent dissociation is another name for acid strength. Large percent dissociations mean stronger acids. Dilutions of the initial concentration of acid increases percent dissociation. This is where the initial concentration of the acid and the hydronium cation concentration both decrease.

Hydroxide salts such as sodium hydroxide and lithium hydroxide are considered to be strong bases. These compounds dissociate completely in water. Bases do not always have to contain hydroxide anions. They can be species where a proton is removed from water and then a hydroxide anion is produced. The value K_b refers to a reaction where a base reacts with water to produce a conjugate acid and a hydroxide anion. Bases other than hydroxides have K_b values less than 1 and are considered to be weak bases. Polyprotic acids have more than one acidic proton, and they dissociate in a step-wise fashion with a K_a value at each step. Weak polyprotic acids are as follows $K_{a1} > K_{a2} > K_{a3}$.

Salts can have neutral, acidic, or basic properties when they are dissolved in water. Salts that have cations of strong bases and anions of strong acids give neutral aqueous solutions. Basic solutions are produced when the acid has a neutral cation, and an anion that is the conjugate base of a weak acid. Acidic solutions are produced when the salt has a neutral anion and a cation that is the conjugate acid of a weak base. Acidic solutions can also be produced by salts that have a highly charged metal cation. Cations such as Fe^{3+} or Al^{3+} are considered to be Lewis Acids with empty p orbitals.

The H-O-X group can act as either an acid or as a base. The O-X portion of the substance is a strong covalent bond if the substance acts as an acid. This is where X is highly electronegative. The O-H bond becomes weak and polarized. Low electronegativities of X cause the O-X bond to be ionic. Hydroxide anions form when the substances are dissolved in water. The Lewis acid-base model describes acid-base behavior based on electron pairs. Lewis acids are electron-pair acceptors, whereas Lewis bases are electron-pair donors. The Lewis model for acids and bases includes the Bronsted model for acids and bases. However, the reverse process does not work.

Chapter 16
Aqueous Equilibria

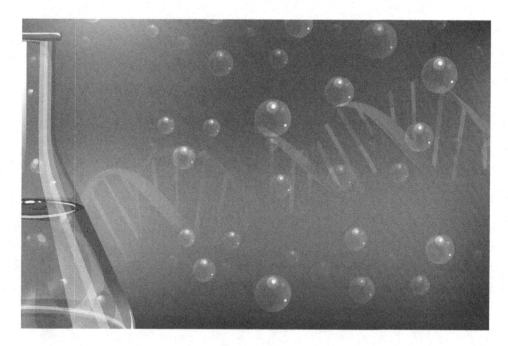

An Experiment

We will discuss solutions involving the weak acid HA and its salt NaA. If we had a solution containing the weak acid HF and its salt NaF, we would be interested to know what chemistry takes place. Dissolving a salt such as NaF produces the cation Na^+ and the anion F^- in aqueous solution. The ions are completely broken apart into its cation and anion. It is considered to be a strong electrolyte. Hydrofluoric acid is considered to be a weak acid, and it is slightly dissociated. The major species in solution are HF, Na^+, F^-, and H_2O. The common ion in the solution is F^-. It is made from both hydrofluoric acid and sodium fluoride. The dissolved sodium fluoride has a certain effect on the dissociation equilibrium on hydrofluoric acid.

Let us compare hydrofluoric acid in two different solutions. Consider the first reaction where we have hydrofluoric acid in aqueous solution, and a second reaction containing hydrofluoric acid and the salt sodium fluoride. Adding sodium fluoride on the right hand side drives the equilibrium to the left by adding fluoride anions. Hydrofluoric acid will dissolve a lot less than when sodium fluoride is added. Fewer hydronium cations are present in solution. Shifting the equilibrium position by adding a common ion in solution such as the fluoride anion is called the common ion effect. Both sodium fluoride and hydrofluoric acid make the solution less acidic than hydrofluoric acid by itself. The common ion effect also takes place in solutions containing polyprotic acids. Protons produced in the first dissociation step inhibits future dissociating steps which also produce protons.

Buffered Solutions

Acid-base solutions containing a common ion is important when we talk about buffering. A buffered solution resist changes in pH whenever hydroxide or hydronium ions are added. Blood is an example of a buffered solution which has the potential to absorb acids or bases during biological reactions without a change in its pH. The pH of blood needs to be constant because our cells only survive in an extremely narrow pH range. Buffered solutions must contain a weak acid and its salt or a weak base and its salt. Using the right components, a buffered solution can be prepared at any pH.

Consider a buffered solution containing a weak acid and its conjugate base where hydroxide anions in aqueous solutions deprotonates the weak acid

$$OH^- + HA \longrightarrow A^- + H_2O$$

The hydroxide anions do not accumulate. The hydroxide anions are replaced by A^- ions. The dissociation reaction for HA is as follows

$$HA \rightleftharpoons H^+ + A^-$$

The equilibrium expression can be written as

$$K_a = [H^+][A^-]/[HA]$$

Rearranging the equation gives us

$$[H^+] = K_a[HA]/[A^-]$$

The pH of a solution is determined by the ratio [HA]/[A⁻] in a buffered solution. Adding hydroxide anions cause the HA to be converted to A⁻. The ratio [HA]/[A⁻] decreases. If the concentrations of HA and A- are present in large initial amounts compared to the hydroxide anion, the ratio [HA]/[A-] will be considered to be small. Since this change is considerably small, the pH stays essentially constant.

Let us take a look at what happens when protons are added to a buffered solution containing a weak acid and the salt of its conjugate base. The A^- anion has a strong affinity for H^+. These added H^+ ions react with A^- to make a weak acid. The following reaction can be overgeneralized as followed

$$H^+ + A^- \longrightarrow HA$$

Free hydronium cations do not accumulate in solution. A net change of A^- to HA takes place. If both of these concentrations of A^- and HA are large compared with the addition of hydronium ions added, very small changes in pH takes place. If we know the K_a value and concentrations of HA and A^-, we can easily calculate the concentration of hydronium cations.

Carrying out some algebraic steps we have the following

$$[H^+] = K_a[HA]/[A^-]$$

$$-\log[H^+] = -\log(K_a) - \log([HA]/[A^-])$$

$$pH = pK_a - \log([HA]/[A^-])$$

$$pH = pK_a + \log([A^-]/[HA])$$

The equation can be stated more simply as

$$pH = pK_a + ([base]/[acid])$$

This is called the Henderson-Hasselbach Equation.

A buffered solution can also be made from a weak base and its conjugate acid. The weak base represented as B can react with the addition of hydronium cations. The following reaction can be represented as

$$B + H^+ \longrightarrow BH^+$$

This product BH^+ is considered to be the conjugate acid. The conjugate acid BH^+ can react with the addition of hydroxide anions to form a base and water. The following reaction can be represented as

$$BH^+ + OH^- \longrightarrow B + H_2O$$

Twenty-First Century Advanced Chemistry

Buffer Capacity

A buffering capacity of buffered solutions shows the amount of protons or hydroxide anions a buffer absorbs without barely changing its pH. Buffers that have large capacities have large concentrations of buffering components. They can absorb lots of protons or hydroxide anions, but still resist changes in pH by small amounts. The ratio [A-]/[HA] determines the pH of a buffered solution. The magnitudes of [HA] and [A-] determines the capacity of a buffered solution.

Suppose we wanted to find the ratio that gives optimal buffering. Consider a buffer solution that contains a huge concentration of an anion and only a small concentration of the original compound with the hydrogen attached. Adding protons to the anion will form the specific compound and produce a large percent change in the compound. This will then produce a large change in the ratio [A-]/[HA]. If hydroxide anions are added to remove protons from the original compound, then the percent change in the concentration of the compound is large. We can get the same results if the initial concentration of the compound is large and the anion is small.

Large changes in [A-]/[HA] will give large changes in the pH value. We should do our best to avoid this stage if we want to have the best buffering capacity. Optimal buffering happens when [HA] equals the [A-]. The ratio [A-]/[HA] is resistant to changes when the hydronium cation or hydroxide anion is added to a buffered solution. Whenever we choose components for the buffer, we need to have the [A-]/[HA] = 1.

$$pH = pK_a + log([A-]/[HA])$$

$$pH = pK_a + log(1)$$

$$pH = pK_a$$

The weak acid's pK_a should be close to the desired pH in the buffer.

Titration Curves

Titration curves are used to determine how much acid or base is in solution. The titrant is the solution of known concentration. It is delivered from a buret into a solution that is to be determined. This is the unknown solution that is going to be determined as it is being consumed. A change in color of the solution is carried out by an indicator that gives the equivalence or stoichoimetric point. Let us take a look at the pH changes that take place in an acid-base titration. Acid-base titrations are monitored by a pH curve or titration curve. The pH values are plotted on the y-axis and the amount of titrant being added is plotted on the x-axis.

The final net ionic equation for strong acid-base titrations is the hydronium cation mixed with the hydroxide anion to give liquid water. If we want to determine the concentration of the hydronium cation at a specific point in the titration, we have to determine how much hydronium cation remains at that specific point and then divide by the total volume of solution. Titration experiments involve small units such as milliliters. When we want to calculate quantities involving small units, we should allow other quantities expressed to be small too. The quantity used to express amount in titrations is millimoles. The millimole is a thousandth of a mole. Molarity was defined as moles of solute per liter of solution. If we divide the numerator and denominator by 1000, we get the ratio millimole of solute per milliliter of solution. An example of a strong acid is nitric acid, and a strong base is sodium hydroxide.

Weak acid-strong base titrations are different from other titrations. In order to calculate the hydronium cation concentration, we have to deal with a weak acid dissociation equilibrium after a certain amount of strong base is added. Carrying out these calculations is similar to doing buffer problems. An example of a weak acid is acetic acid, and an example of a strong base is sodium hydroxide. Weak base-strong acid titrations are also different from other titrations. There is also an equilibrium for this type of acid-base titration. An example of a weak base is ammonia, and an example of a strong acid is hydrochloric acid.

Acid-Base Indicators

The equivalence point for an acid-base titration can be found by using a pH meter. pH meter devices monitor the pH and plots the titration curve. The equivalence point is considered to be the center vertical region of a pH curve. The equivalence point for an acid-base titration can also be found using an acid-base indicator. These indicators are used to mark the end point during the course of a titration by changing color. Note that the equivalent point does not always have to be the same as the end point. Remember the end point is where the indicator changes color. Indicators should be selected to make sure the error does not make a difference.

Acid-base indicators are weak acids, and we can represent them as HIn. When the proton is attached, they give one color. But when the proton is detached, they give a different color. An example of an indicator is phenolphthalein. It is colorless in its HIn form and pink in its In⁻ form.

Let us take a look at the equilibrium expression

$$HIn(aq) \rightleftharpoons H^+(aq) + In^-(aq)$$

The color compound HIn is red whereas the anion In⁻ is blue. The acidity constant can be calculated as

$$K_a = [H^+][In^-]/[HIn]$$

Rearranging the equation gives

$$K_a/[H^+] = [In^-]/[HIn]$$

The more hydroxide anion that is added to the solution, the concentration of the hydronium ion decreases. The equilibrium shifts itself to the right. The concentration of HIn changes to In⁻. There will be a point in the titration when enough of the In⁻ form is present in the solution. A slight purple tint will appear. There will be a color change from red to slight purple. Most indicators require approximately a tenth of the initial form to be converted to the other form before the new color appears. When we are titrating an acid with a base, a color change occurs for the pH at

$$[In^-]/[HIn] = 1/10$$

This is the ratio for the titration of an acid. We can use the Henderson-Hasselbach Equation for determining the pH for an indicator changing color. The equation then becomes

$$pH = pK_a + \log([In^-]/[HIn])$$

The K_a value is the dissociation constant for the indicator HIn. A color change is visible when

$$[In^-]/[HIn] = 1/10$$

The pH at which the color change occurs is

$$pH = pK_a + \log(1/10)$$

$$pH = pK_a - 1$$

When a basic solution is going to be titrated, the indicator HIn will begin to exist as In⁻. When an acid is added, more of the indicator HIn will form. There will be a visible color change at 10 parts In⁻ to 1 part HIn. We can write the ratio as

$$[In^-]/[HIn] = 10/1$$

This is the ratio for the titration of a base. If we substitute this ratio into the Henderson-Hasselbach Equation, we get

$$pH = pK_a + \log(10/1)$$

$$pH = pK_a + 1$$

Acid-base indicators that have dissociation constant K_a have a transition of color over the pH range of $pK_a \pm 1$. When choosing an indicator for a titration, the end point for the indicator and the equivalence point should be as close to possible. It is easier to choose an indicator if the pH range is large and is near the equivalence point. A change in pH produces a sharp end point. There is a complete change in color that occurs when one drop of titrant is added. Changes of color for indicators will be sharp, and in addition there is a wide range of indicators to choose from. Most results should be within one drop of titrant using specific indicators suitable for the chemical reaction. It is best to choose an indicator whose pH range has a midpoint that is close to the pH of the equivalence point.

Solubility Equilibria

Let us consider equilibria that has to do with solids dissolving to form aqueous solutions. When an ionic solid dissolves in water, it separates into a cation and an anion. An example is magnesium fluoride dissolved in water which produces the magnesium cation and two moles of fluoride. When magnesium fluoride is first added to water, none of the ions are present. As the dissolving process proceeds, the concentrations of magnesium and fluoride increase. This makes it more likely that the ions collide and reform the solid. We have two competing reactions. The dissolution and the reverse reaction takes place. Dynamic equilibrium is achieved. No more of the solid dissolves. We say that the solution is saturated. The solubility product constant, K_{sp}, can be calculated for the equilibrium process.

The compound magnesium fluoride is a solid so it is not part of the equilibrium expression. The excess amount of solid does not affect the position of the solubility equilibrium. Doubling the surface area of the solid doubles the rate of dissolving which in turn doubles the rate of reformation. An excess amount of solid does not have an effect on the position of the equilibrium. Grinding the solid or stirring the solution neither changes the amount of solid dissolved at equilibrium. Amount of excess solid or the size of the particles does not shift the position of the solubility equilibrium.

Solubility product is considered to be an equilibrium constant. It has only one value for a solid at a specific temperature. However, solubility is considered to be an equilibrium position. A salt mixed in with water has a specific temperature and a particular solubility. If there is a common ion present in solution, solubility varies with concentration of the common ion. The K_{sp} value gives information about solubility. Relative solubilities only work by comparing K_{sp} values for salts that have the same total number of ions. When water contains a specific ion with the dissolving salt, we have the common ion effect. The common ion effect has to do with the solubility of a solid being lowered if the solution contain ions common to the solid. The pH for a solution can also affect the solubility of a salt. If an anion such as X^- is a good base, the salt MX shows increased solubility in an acidic solution.

Analyzing Solutions

When we mix solutions, different kinds of reactions occur. Let us take a look at how to predict whether a precipitate forms when two solutions are mixed. The ion product is used for equilibrium concentrations. The criteria whether a precipitation will take place is when the ion product is greater than the solubility product constant, then a precipitation takes place and continues until the concentrations get reduced until the solubility product is satisfied. If the ion product constant is less than the solubility product constant, there is no precipitation. Metal cations mixed with solutions can take place too. Metal ions mixed with aqueous solutions can be separated by selective precipitation. This can be carried out by using a reagent that contains an anion and forms a precipitate with one or a few of the metal ions in the mixture.

We also find there can be complex ions that are charged species in solution. Complex ions are considered to be charged species that consist of a metal ion being surrounded by ligands. Ligands are Lewis bases. Recall that Lewis bases have lone electron pair that is denoted to an empty orbital on the metal ion which then makes a covalent bond. The coordination number is the number of ligands attached to the metal ion. Metals ions tend to add ligands one by one in steps. These steps are characterized by equilibrium constants. The equilibrium constants are also called stability constants or formation constants.

Some ionic solids are nearly water-insoluble and somehow they must be dissolved in aqueous solutions. When qualitative analysis groups get precipitated, the precipitates redissolve to separate the ions within each group. Once the precipitate that is mixed becomes separated from the solution, it is redissolved to identify which cations there are. Two strategies for dissolving a water insoluble ionic solid can be carried out. The anion of a solid is a good base. Its solubility is increased by adding acid to the solution. When the anion is not basic enough, the ionic solid can be dissolved in solution that contains a ligand and forms stable complex ions with the cation. Some solids can be so insoluble that combining reactions are needed to dissolve them. The solubility of many kinds of salts increase with temperature. Heating can make a salt soluble enough.

Conclusion

A weak acid and its salt mixed together results in a buffering effect. Buffered solutions resist changes in pH when hydroxide ions or hydronium ions are added to the mixture. Buffers contain a weak acid and its salt or a weak base and its salt. Buffered solutions have to do with the pH being governed by the ratio of the concentration of the anion to the concentration of the acidic species. The Henderson-Hasselbach Equation can be used to calculate the pH of a solution. A buffer's capacity depends on the amounts of the weak acid and its conjugate base. Buffers take place because large amounts of the weak acid and the conjugate base cause the ratio of the concentration of the conjugate base to its acid to stay constant when either hydronium or hydroxide ions are added. Best buffering results when the concentration of the weak acid equals the concentration of the conjugate base.

An acid-base titration curve is monitored by plotting pH of the solution versus the volume of titrant added. Strong acid-strong base titrations have sharp changes in pH near the equivalence point. The equivalence point or stoichiometric point is when the original acid or base is exactly consumed by the titrant base or acid. Weak acids titrated with strong bases have different titration curves. The curve rises in a sharp manner and then levels off due to buffering effects of weak acid and its salt mixture. For a titration involving a weak acid with a strong base, the pH at the equivalence point is greater than seven. The basic nature of the conjugate base of the weak acid has a stronger effect. For a titration involving a weak base with a strong acid, the pH at the equivalence point is less than seven. The acidity of the conjugate acid of the weak base has a stronger effect. For titrations involving strong acid-strong base, the pH at the equivalence point should be seven. Equivalence points are determined by stoichiometry and not by pH.

Acid-base titrations can also be followed by using indicators. Indicators mark the end point by a change in color. They are weak acids that have the form HIn and In⁻. These show different colors. Most indicators have ratios of the minor form to the major form to be 1/10 before a color change takes place. Indicators are best chosen to have the equivalence point and the end point coincide. Equilibrium is also applied when solid dissolves in water. The solubility product, K_{sp}, is an equilibrium constant. Solubility has to do with equilibrium position. For a salt, the K_{sp} value is determined by measuring its solubility. The solubility of a salt can be found if the K_{sp} value is already known. Adding a common ion to a salt pushes the dissociation equilibrium to the left. This reduces the solubility of a salt. This is called the common ion effect. Salts that are slightly soluble in pure water can be increased as solubility equilibrium is shifted to the right by lowering the concentrations of the metal cation and anion. Suppose the anion is a good base, adding hydronium cations forms an acidic species and lowers the anion concentration. This increases the salt's solubility. The concentration of the metal cation is lowered by the addition of ligands to make complex ions. Precipitates can be predicted when two solutions are mixed and by using the ion product. Initial concentrations are used for the equation for the ion product constant. If the ion product is greater than the solubility product constant, salts get precipitated. Mixtures involving ions are separated by selective precipitation. This procedure involves adding reagents to precipitate single or group of ions.

Chapter 17

Thermodynamics

Plasma Ball

The First Law of Thermodynamics is another way stating the Law of Conservation of Energy. Energy is neither created or destroyed. The energy of the universe remains constant. Even though energy remains the same, energy can be changed into different forms based on physical or chemical processes. We can think of the First Law of Thermodynamics as a way to keep track of energy. We like to know how much energy does the change involve, the flow of energy into and out of the system, and the final form of the energy. The problem with the First Law of Thermodynamics is that it does not tell us why a process occurs in a specific direction.

Spontaneous processes happen on its own without any outside disturbance. Spontaneous processes can either be fast or slow. Thermodynamics tells us the direction a process occurs, but it does not say anything about the speed of the process. Reaction rates depend on the energy of activation, concentration, temperature, and catalysts. These processes can be explained using the simple collision model. When we describe a reaction, chemical kinetics focuses on the pathway for reactants and products. Thermodynamics, however, only deals with the initial and final states. The pathway between reactants and products is not important.

Thermodynamics predicts whether a process occurs, but it does not give information about the time that is required for the process to take place. Both thermodynamics and kinetics are needed to describe reactions in greater detail. Processes that tend to be spontaneous in only one direction has to do with an increase in a property called entropy. We can denote this by the symbol S. For spontaneous processes, the driving force has to do with an increase in entropy of the universe. Entropy has to do with the measure of randomness or disorder. Processes tend to go from a lower entropy to a higher entropy. Entropy is considered to be a thermodynamic function. It describes the positions or energy levels for a system in a given state.

The more ways a state is achieved, the greater the probability of finding the state. Nature has its way to proceed toward states with higher probabilities. For example, the driving force of an ideal gas into a vacuum is a spontaneous process. The driving force has to do with probability. There are more ways the gas spreads throughout the container than there are ways to be in a specific state. The gas goes through a spontaneous process and gives a uniform distribution. The probability of a particular arrangement or state depends on the number of ways or micro states in which the arrangement is achieved.

A gas placed in one end of the container can expand and fill the entire vessel. This can happen because there is a large number of gas molecules which also gives us huge number of micro states for equal numbers of molecules at both ends. The opposite process is not probable since only one micro state can lead to one arrangement. The process does not occur spontaneously. This is called positional probability since it depends on the number of configurations in positioned micro states that gives a specific state. Gases expand into a vacuum to give a uniform distribution since the expanded state has the highest positional probability or the largest entropy of the states available to the system.

Positional probability can also be illustrated by changes of state. Positional probability tends to increase from a solid to a liquid to a gas. Substances have a smaller volume in the solid state than in the gaseous state. Molecules are close together in the solid state. Not that many open positions are available in the solid state. Molecules in the gaseous state are far apart. There are more open positions available for them. We can think of the liquid state being closer to the solid state than the gaseous state. We can think of the trend $S_{solid} < S_{liquid} < S_{gas}$. Mixing has to do with the increase volume available to particles of each component of the mixture. When we mix two liquids, molecules from each liquid will have more available volume and therefore more available positions.

Positional entropy is also considered in the formation of solutions. Solutions are also formed and favored due to the natural tendency for substances to mix. The entropy change is positive for two pure substances to be mixed. Entropy increases because there are many more micro states when mixed than for separate conditions. The effect is due to the increased volume to a given particle after mixing takes place. Mixing two liquids to form a solution allows the molecules to have more volume and more available positions. Positional entropy increases, and it has to do with mixing that favors solutions being formed.

Entropy

Processes tend to be spontaneous when they result in an increase in disorder. Nature tends to move to the most probable state. This principle can be stated in terms of entropy. In a spontaneous process, there is an increase in entropy of the universe. We consider this as the Second Law of Thermodynamics. Remember the First Law of Thermodynamics tells us the energy of the universe is constant. Energy is considered to be conserved in the universe. However, entropy is not conserved. The Second Law of Thermodynamics can be stated as the entropy of the universe is increasing. The universe can be divided into a system and the surroundings. The change in entropy of the universe is

$$\Delta S_{univ} = \Delta S_{sys} + \Delta S_{surr}$$

This is where ΔS_{sys} and ΔS_{surr} represent changes in entropy for the system and the surroundings. In order to predict whether a process will be spontaneous or not, the sign of ΔS_{univ} should be known. If ΔS_{univ} is positive, then the entropy of the universe increases. The process is spontaneous according to the direction it is written. If ΔS_{univ} is negative, then the process becomes spontaneous in the other or opposite direction. If ΔS_{univ} is zero, then the process does not have a tendency to occur. The system stays at equilibrium. Predicting if a process is spontaneous or not, we should consider entropy changes that happen in the system, the surroundings, and their sum.

The Second Law of Thermodynamics states that a process will be spontaneous if the entropy of the universe increases when a process occurs. For a process at constant temperature and pressure, the change in free energy of the system is used to predict the sign of ΔS_{univ} and the direction in which it is spontaneous. These ideas are applied to changes of state and solution formation. We are interested to apply the Second Law of Thermodynamics to reactions.

Let us consider entropy changes that accompany chemical reactions under conditions of constant temperature and pressure. Entropy changes in the surroundings are determined by the flow of heat that happens when the reaction takes place. Entropy changes in the system can be determined by positional probability. For example, if we were to take Reactant A plus Reactant B we would get Reactant C based on a combination reaction. Lowering the moles of products in the system leads to less positional disorder. Fewer molecules in the product for this example mean fewer possible configurations. On the other hand, if we had one molecule and it became two molecules by decomposition, we would find positional entropy increases. Reactions involving gaseous molecules has to do with the change in positional entropy being dominated by the number of gaseous molecules in the reactants and products. If there are more gaseous molecules in the product than in the reactant, positional entropy tends to increase, and the change in entropy will be positive.

When we talk about thermodynamics, we are interested in the change in a certain function. The change in enthalpy determines whether the reaction is exothermic or endothermic at constant pressure. Changes in free energy determines if a process is spontaneous at constant temperature and pressure. Changes in thermodynamic functions are sufficient. Absolute values for many thermodynamic functions such as enthalpy or free energy cannot be determined. But we can assign absolute value entropy values. A perfect crystal at 0 Kelvins has the internal arrangement absolutely regular. To achieve this disorder, every particle must be in its place. The entropy of a perfect crystal at 0 K is considered to be zero. This is considered to be the Third Law of Thermodynamics.

Increasing the temperature of a perfect crystal causes an increase in the random vibrational motions. Disorder also increases within the crystal. The entropy of a substance tends to increase with temperature. The entropy is zero for crystals at 0 K. Entropy values for a substance at a specific temperature is calculated by knowing how temperature depends on entropy. Entropies of substances increase going from solid to liquid to gas. Entropy is a state function. It does not depend on the pathway. The entropy change of a reaction can be calculated by taking the difference between standard entropy values of products and reactants. The formula is as follows

$$\Delta S^\circ_{reaction} = \sum n_p S^\circ_{products} - \sum n_r S^\circ_{reactants}$$

Entropy is considered to be an extensive property. Entropy depends on the amount of substance present. The number of moles of a given reactant or product must be taken into account.

Free Energy

The value of ΔG can be calculated for chemical reactions under different conditions. Systems achieve the lowest free energy by reaching equilibrium instead of going to completion. The value of ΔG for a reaction system indicates whether the products or reactants are favored under a given set of conditions. It does not mean the system proceeds to pure products if the value of ΔG is negative or should it remain at pure reactants if the value of ΔG is positive. The system spontaneously goes to the equilibrium position. Again, this is where we have the lowest possible free energy that is available. The value of ΔG° indicates where the position will be at.

The quantity ΔS_{univ} is used to predict the spontaneity for a process. Free energy is also related to spontaneity. Free energy is symbolized by the letter G. It is given by the equation

$$G = H - TS$$

This is where H is the enthalpy, T is the temperature in Kelvins, and S is the entropy. Processes that occur at constant temperature have free energy changes given by the equation

$$\Delta G = \Delta H - T\Delta S$$

All of the quantities here refer to the system. Dividing by -T gives us

$$\Delta G/\text{-}T = \Delta H/\text{-}T + \Delta S$$

At constant temperature and pressure,

$$\Delta S_{surr} = \Delta H/\text{-}T$$

$$\Delta G/\text{-}T = \Delta H/\text{-}T + \Delta S$$

$$\Delta S_{univ} = \Delta S_{sys} + \Delta S_{surr}$$

This means for a process to be spontaneous at constant temperature and pressure, ΔG will have to be negative. Free energy decreases when ΔS_{univ} is positive. Let us look at specific cases when the change in entropy or the change in enthalpy are either positive or negative. If the change in entropy and the change in enthalpy is negative, then the result is spontaneous at all temperatures. If the change in entropy is positive and the change in enthalpy is positive, the result is spontaneous at high temperatures. If the change in entropy is negative and the change in enthalpy is negative, the result is spontaneous at low temperatures. If the change in entropy is negative and the change in enthalpy is positive, the result is not spontaneous at any temperature.

The standard free energy change, symbolized as $\Delta G°$, is the change in free energy that occurs if the reactants in their standard states are converted to products in their standard states. The free energy change for a reaction cannot be measured directly. There is no instrument that measures $\Delta G°$ for a reaction. It is calculated from other measured quantities. The free energy values do not tell us about reaction rates, but it does tell us about its equilibrium position. Knowing free energy values for many reactions allows us to compare the tendency of reactions to occur. The more negative the free energy value, the further the reaction will go to the right to reach equilibrium. Standard state free energies must be used for comparison since free energy varies with pressure or concentration.

If we are to measure everything accurately, all reactions should be compared under the same concentration or pressure. The standard free energy change is normally calculated by the equation $\Delta G° = \Delta H° - T\Delta S°$. A second method for calculating free energy are procedures similar for finding the change in enthalpy values using Hess's Law. We should note that free energy is an extensive property. A third method for calculating free energy change uses standard free energies of formation. The standard free energy of formation is the change in free energy that accompanies the formation of 1 mole of the substance from its elements with all reactants and products in their standard states. The free energy changes for specific reactions can be calculated using the equation

$$\Delta G^{\circ}_{reaction} = \Sigma n_p \Delta G_f^{\circ}(products) - \Sigma n_r \Delta G_f^{\circ}(reactants).$$

When using this equation, the standard free energy of formation of an element in its standard state is zero. The number of moles of each reactant and product is taken into consideration when calculating ΔG° for a reaction.

Systems at constant temperature and pressure can proceed spontaneously in a direction that ends up lowering the free energy. This explains why reactions proceed until they reach equilibrium. The equilibrium position is the lowest free energy value for a reaction system. The free energy of the reaction system changes as a reaction goes further. Free energy depends on the pressure of a gas or on the concentration of the particular species in solution. Let us take a look at how pressure depends on the free energy of a gas. Pressure affects thermodynamic functions that consist of free energy, enthalpy, entropy, and temperature.

When we are dealing with ideal gases, enthalpy does not depend on pressure. Entropy depends on pressure because pressure depends on volume. If we were to consider one mole of an ideal gas at a specific temperature and a volume of about five liters for example, we find the gas has more positions available for molecules than a volume of about a liter. Positional entropy is greater for a larger volume. At a given temperature for a mole of ideal gas, we find $S_{large\ volume} > S_{small\ volume}$. Pressure and volume are inversely related as $S_{low\ pressure} > S_{high\ pressure}$. The entropy and free energy for an ideal gas depend on pressure. We can show that

$$G = G^{\circ} + RTln(P)$$

where G° is the free energy of the gas at a pressure of 1 atm, G is the free energy of the gas at a pressure of P atm, R is the universal gas constant, and T is the Kelvin temperature.

Equilibrium

Let us take a look at how equilibrium relates to free energy. When mixing components together for a given chemical reaction, they can proceed rapidly or slowly which in turn depends on kinetics. The components proceed to the equilibrium position. The equilibrium position is the point where the forward and reverse reaction rates become equal. The equilibrium point occurs at the lowest value for free energy available for a reaction system. If we consider

$$A(g) \rightleftharpoons B(g)$$

where 1 mole of gas A is placed in a reaction vessel, gas A reacts with itself to form B. The total energy for the system changes. As gas A changes to gas B, the free energy for A decreases because the pressure is decreasing.

On the other hand, the free energy of B will increase because the pressure of B will increase. The reaction proceeds to the right as long as the total free energy for the system decreases. At a specific point, the pressures of A and B make the free energy of A equal to the free energy of B. The system has reached equilibrium. The value for ΔG equals zero for gas molecule A changing to gas molecule B. We say the system has reached the minimum amount of free energy. There is no more driving force to change A to B or from B to A. The system stays the same at this position. The pressures for A and B remain constant. When substances go through a chemical reaction, the reaction proceeds to the lowest free energy or equilibrium which corresponds to

$$G_{products} = G_{reactants}$$

The difference of these two G's would be

$$\Delta G = G_{products} - G_{reactants} = 0$$

Let us take a look at a quantitative relationship between the free energy and the equilibrium constant. We have the equation

$$\Delta G = \Delta G° + RTln(Q)$$

At equilibrium, the value of ΔG equals zero and Q equals K. Substituting $\Delta G = 0$, we have the equation

$$0 = \Delta G° + RTln(K)$$

The equation can be rewritten as

$$\Delta G° = - RTln(K)$$

If $\Delta G° = 0$, then the free energies of the reactants and products become equal when the components are in the standard states. The system is at equilibrium when the pressures of the reactants and products are 1 atmosphere. This means that K equals 1. If $\Delta G° < 0$, then this means $\Delta G°$ is negative. The value for $G°_{products}$ is less than $G°_{reactants}$. Suppose the flask contains reactants and products, the system will not be at equilibrium. The system will also adjust itself to the right to reach equilibrium. The value of K will be greater than 1 since at equilibrium the pressure of the products is greater than 1 atm. The pressure of the reactants will be less than 1 atm at equilibrium. If $\Delta G° > 0$, this means $\Delta G°$ is positive. If the flask which contains reactants and products that are all at 1 atm, the system does not need to be at equilibrium. The system adjusts itself to the left where the reactants are. This is where they have a lower free energy to reach equilibrium. The value of K will be less than 1. At equilibrium, the pressures of the reactants will be greater than 1 atm. The pressure of the products will be less than 1 atm. Temperature that depends on the equilibrium constant include

$$\Delta G° = - RT\ln(K)$$

$$\Delta G° = \Delta H° - T\Delta S°$$

$$\ln(K) = - \Delta H°/RT + \Delta S°/R$$

$$\ln(K) = - (\Delta H°/R)(1/T) + \Delta S°/R$$

Work

Physical and chemical processes are important because we want to use them to do work. At constant temperature and pressure, the free energy change whether it is positive or negative tells us whether the process is spontaneous or not. This helps us to avoid wasting effort for processes that have no tendency to occur. A chemical reaction can be thermodynamically favored but may not occur to a certain extent because it may be slow. We would need to find a catalyst to speed up the reaction. If the reaction is prevented from happening based on its thermodynamic characteristics, we could be wasting time looking for a catalyst.

The free energy change is important since it tells us how much work could be done for a process. Maximum possible useful work can be obtained from a constant temperature and pressure process which is equal to the free energy change. The equation can be rewritten as

$$w_{max} = \Delta G$$

The value of ΔG for a spontaneous process gives us the energy that is free to do useful work. For nonspontaneous processes, the ΔG value indicates the minimum amount of work that is needed to expend the process to occur. The value of ΔG tells us how close the process is to 100% efficiency. The work we obtain from a spontaneous process is less than the maximum possible amount. Obtaining the maximum amount of work from a spontaneous process is only hypothetical. Real pathways waste energy. Real processes are also considered to be irreversible. In real cyclic processes, work changes to heat in the surroundings. The entropy of the universe ends up increasing. This is how we can state the Second Law of Thermodynamics. Thermodynamics indicate work potential for a process and indicates we can never reach this potential.

Conclusion

The First Law of Thermodynamics states energy of the universe is constant. But this does not give us information why a process occurs in a specific direction. Spontaneous processes occur without an outside convention. Entropy is the driving force for a spontaneous process. Entropy is thought of as a measure of randomness or disorder. It describes the positions or energy levels for a system in a given state. Nature tends toward states that have the highest probabilities taking place. The Second Law of Thermodynamics states that for a spontaneous process there is an increase in the entropy of the universe. Based on entropy, thermodynamics predict the direction a process occurs. But it does not predict the rate at which a process occurs. This is carried out using the principles of kinetics.

A process can be considered to be spontaneous by considering the entropy changes that take place in the system and surroundings. This equation can be expressed as $\Delta S_{univ} = \Delta S_{sys} + \Delta S_{surr}$. At constant pressure and temperature, the change in entropy of the system is dominated by the changes taking place in positional entropy. The change in entropy of the surroundings is determined by the amount of heat flow. The equation can be expressed as $\Delta S_{surr} = -(\Delta H/T)$. The sign of the change based on the entropy in the surroundings depend which way heat is flowing. When the positional entropy of the surroundings is positive, we have an exothermic process. When the change in positional entropy of the surroundings is negative, we have an endothermic process. The magnitude of the change in entropy of the surroundings depend how much energy flows as heat and the temperature that is transferred.

Entropy changes for a system is dominated by the change in gas molecules. If there are less gas molecules on the product side, this mean there is a decrease in entropy. Complex molecules have higher standard entropies than a simple molecule. The Third Law of Thermodynamics states the entropy of a perfect crystal at a temperature of 0 K is zero. At temperatures above 0 K, molecules tend to vibrate and disorder takes place. Free energy is defined as the equation $G = H - TS$. Processes occurring at constant pressure and temperature is spontaneous in the direction where the free energy decreases.

The standard free energy change is represented by $\Delta G°$. It is the change in free energy that occurs if the reactants in their standard states get converted to the products in their standard states. Free energy is a state function. The equation can be represented as $\Delta G°_{reaction} = \Sigma n_p \Delta G_f°(products) - \Sigma n_r \Delta G_f°(reactants)$ where ΔG_f is the standard free energy of formation for a substance. In other words, $\Delta G°$ is the free energy change that accompanies the formation for a substance from its constituent elements with the reactants and products in their standard states. The standard free energy of formation of an element in the standard state is zero.

Free energy can also depend on pressure. This can be represented as $G = G° + RT\ln(P)$. The ΔG value can be calculated from the equation $\Delta G = \Delta G° + RT\ln(Q)$ where Q is the reaction quotient. Equilibrium occurs where the free energy is at the minimum value for the system. The relationship between the standard free energy change $\Delta G°$ and the equilibrium constant is $\Delta G° = -RT\ln(K)$. When $\Delta G°$ equals zero, the system is at equilibrium. The products and reactants are in their standard states when the equilibrium constant is one. When $\Delta G°$ is less than zero, $G°_{products} < G°_{reactants}$ the system will then adjust to the right in order to obtain equilibrium. This is where the equilibrium constant is greater than one. When $\Delta G°$ is greater than zero, $G°_{reactants} < G°_{products}$, the system will then shift to the left to obtain equilibrium. This is where the equilibrium constant is less than one. The maximum work obtained is equal to the change in free energy. This can be written as w_{max} is equal to ΔG. For real processes, the actual work is less than the maximum work. The total energy of the universe is constant when energy is used to do work. But it is not very useful. After it is used to do work, energy that is concentrated spreads out into the surroundings as thermal energy.

Chapter 18

Electrochemistry

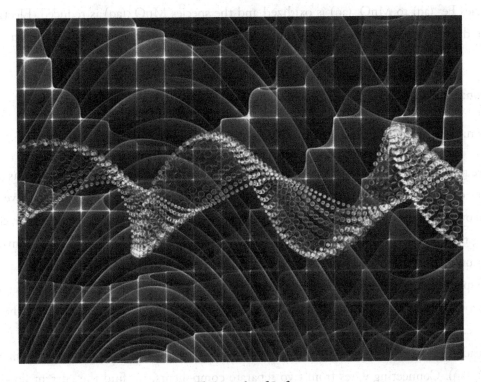

Chemicals of Life

Starting our vehicles, using our calculators, using our digital watches, or even using the radio depends on electrochemical reactions. Our society seems to run on batteries. Small batteries and silicon-chip technology has made it possible to have small calculators, clocks, and tape recorders. Iron, for example, that corrodes is also an electrochemical process. Industrial materials such as chlorine, aluminum, and sodium hydroxide are prepared by carrying out electrolytic processes. Electrodes are used in electrochemistry, and they are specific for a given molecule or ion such as the hydronium cation, the fluoride anion, the chloride anion, and several other ions. Methods such as these are used to trace pollutants in natural waters or in extremely small quantities of chemicals in human blood that could tell us whether a specific disease would develop. We can define electrochemistry as the study of the interchange of chemical and electrical energy. Two processes are involved using oxidation-reduction reactions. Electric current being generated from spontaneous chemical reactions and the use of current to produce chemical change.

Galvanic Cells

Oxidation-reduction reactions or redox reactions involve transferring electrons from a reducing agent to an oxidizing agent. Oxidation involves losing electrons or an increase in oxidation number. Reduction involves gaining electrons or a decrease in oxidation number. Consider the reaction

$$8H^+(aq) + MnO_4^-(aq) + 5Fe^{2+}(aq) \longrightarrow Mn^{2+}(aq) + 5Fe^{3+}(aq) + 4H_2O(l)$$

The species $Fe^{2+}(aq)$ to $MnO_4^-(aq)$ is oxidized and the species $MnO_4^-(aq)$ is reduced. Electrons are transferred from $Fe^{2+}(aq)$ to $MnO_4^-(aq)$. Let us take a look at the half reactions for the reaction given above.

Reduction: $8H^+(aq) + MnO_4^-(aq) + 5e^- \longrightarrow Mn^{2+}(aq) + 4H_2O(l)$

Oxidation: $5[Fe^{2+}(aq) \longrightarrow Fe^{3+}(aq) + e^-]$

When we multiply "5" in the second reaction, what we are doing is showing the reaction has to occur five times each time the first reaction that takes place. Summing both half reactions gives us the balanced overall reaction. When the species $MnO_4^-(aq)$ and $Fe^{2+}(aq)$ are in the same solution, electrons are being transferred when the reactants collide with each other. The chemical energy involved in the reaction gives us no useful work. Instead, heat is released. There has to be a way to harness the energy. Separating the oxidizing agent and the reducing agent and having an electron transfer to occur through a wire can help us harness the energy. Current that is produced in the wire by the flow of electrons can be directed through a device like an electric motor to provide useful work.

Considering the reaction above, electrons should end up flowing through a wire from $Fe^{2+}(aq)$ to $MnO_4^-(aq)$. Connecting wires from two separate components, we find the current flows for an instant and then automatically stops. Current stops flowing because there is charge buildups in the two separate compartments. In order to solve the problem, the solutions should be connected so ions keep flowing. There is a net charge in each compartment that is zero. The connection involves a salt bridge which is a U-tube filled with an electrolyte. A porous disk in a tube connecting the solutions can also be used. Either devices allow ions to flow without the solutions being mixed. Allowing ion flow gives us a complete circuit. Electrons flow by going through the wire from a reducing agent to an oxidizing agent. Ions flow from one compartment to another to keep the net charge zero. Galvanic cells are devices in which chemical energy changes to electrical energy. Galvanic cells use spontaneous redox reactions to give current that is used to do work. The reaction happens at the interface between the electrode and the solution where electron transfer takes place. The electrode compartment where oxidation takes place is called the anode. The electrode compartment where reduction occurs is called the cathode.

Standard Reduction Potentials

Reactions in galvanic cells are oxidation-reduction reactions that are broken down into two half-reactions. If we assign a potential to each half reaction, then we can construct a cell from a given pair of reactions. The overall potential can be obtained by summing the half-cell potentials. Let us consider the reaction

$$2H^+(aq) + Zn(s) \longrightarrow Zn^{2+}(aq) + H_2(g)$$

For this kind of cell, the anode compartment contains a zinc metal electrode. The ions $Zn^{2+}(aq)$ and $SO_4^-(aq)$ are present in aqueous solution. The anode reaction is the oxidation half-reaction. We can represent this as

$$Zn(s) \longrightarrow Zn^{2+}(aq) + 2e^-$$

The Zn metal produces Zn^{2+} cations that go into solution. The zinc metal is giving up electrons which then flow through the wire. We are also going to assume that all cell components are in their standard states. The solution that is in the anode compartment has a concentration of 1 M Zn^{2+}.

The cathode reaction is the reduction half-reaction. In this case, let us look at the reaction

$$2H^+(aq) + 2e^- \longrightarrow H_2(g)$$

The cathode consist of a platinum electrode that is in contact with 1 M H^+ ions is surrounded by hydrogen gas at 1 atm. The platinum electrode is used because it is a chemically inert conductor. For this type of reaction, this kind of electrode is called the standard hydrogen electrode. It is possible to measure the total potential of the cell. There is no way to measure the potentials of individual electrode processes. Obtaining potentials for half-cells, we would have to divide the total cell potential. If we were to assign for the following reaction

$$2H^+(aq) + 2e^- \longrightarrow H_2(g)$$

the concentration of the hydronium cation to be 1 M and the pressure for hydrogen to be 1 atm. The potential would be exactly zero volts. If that is the case, then the reaction

$$Zn \longrightarrow Zn^{2+}(aq) + 2e\text{-}$$

will have the same number of volts as the overall reaction. The overall reaction is

$$\epsilon°cell = \epsilon°_{H+ \to H2} + \epsilon°_{Zn \to Zn2+}$$

where the superscript ° indicates standard states employed. Setting the standard potential for the half reaction

$$2H^+(aq) + 2e^- \longrightarrow H_2(g)$$

equal to zero, we are able to assign values to other half-reactions. This is considered ideal behavior. The $\epsilon°$ values that correspond to reduction half-reactions with solutes at 1 M and gases at 1 atm are called standard reduction potentials.

We can combine two half reactions to give a balanced redox reaction. One of the redox reactions such as the reduction half reaction has to be reversed. The sign of the potential for the reversed half-reaction should also be reversed. The number of electrons lost should equal the number of electrons gained. The half reactions should be multiplied by integers in order to achieve the balanced equation. The value of $\epsilon°$ is not changed when the reaction is multiplied by an integer. Standard reduction potentials are intensive properties. In other words, they do not depend on the number of times a reaction occurs. The potential does not get multiplied by the integer in order to balance the cell reaction.

Line Notation

Line notation is used to describe electrochemical cells. Using this notation, anode components are listed on the left and cathode components are listed on the right. Double vertical lines are separated in the middle to indicate a salt bridge or a porous disk. A phase boundary is indicated by a single vertical line. Let us consider the general case

$$M_1(s) \mid M_1^+(aq) \parallel M_2^+(aq) \mid M_2(s)$$

Vertical lines between M_1 and M_1^+ and between M_2^+ and M_2 are shown as indicated. The substances shown for the anode is on the far left and the substances shown for the cathode is on the far right. If the components are all ions in the redox reaction, then an inert conductor must be used. None of the dissolved ions can serve as an electrode. The electrode to be used will be platinum.

Four items are used to describe a galvanic cell. The items are the cell potential being always positive for a galvanic cell and the balanced reaction, the direction of electron flow that gives a positive ϵ_{cell} value, determining the location of the anode and the cathode, and the kind of electrode and ions present in each compartment. An inert conductor must be used if none of the substances are conducting solids.

Potential, Work, and Energy

Let us now take a look at the relationship between thermodynamics and electrochemistry. The work that is accomplished when electrons are transferred through a wire depend on the thermodynamic force behind the electrons. The electromotive force is defined based on a potential difference in volts between two points in the circuit. A volt represents a joule of work per coulomb of charge transferred. We can express the equation as

$$\text{Emf(V)} = \text{Potential Difference (V)} = \text{Work (J)} / \text{Charge (C)}$$

In other words, 1 joule of work gets produced or is required that depends on the direction when 1 coulomb gets transferred between two points for a circuit that differs by a potential difference of 1 volt. Work can be viewed either as the point of system or the point of surroundings. Work that flows out of a system is indicated by a minus sign. Cells that produce a current have their cell potential be positive. The current is used to do work. Thus, the cell potential and work have opposite signs. We can represent this in equation form as

$$\epsilon = - w/q$$

The maximum work for a cell is obtained at the maximum cell potential. In order to get electrical work, current needs to flow. As current flows, some of the energy gets wasted through frictional heating. The maximum work cannot be obtained. For real spontaneous processes, some of the energy is wasted. The actual work is always less than the calculated maximum. The entropy of the universe always increases in any spontaneous process. The only way maximum work can be obtained is through a hypothetical reversible process. For galvanic cells, this involves an extremely small amount of current flow and a lot of time in order to do work. Although we can never achieve the maximum amount of work through the discharge of a galvanic cell, it is possible to measure the maximum potential. There is an extremely small amount of current flow when a cell potential gets measured with a potentiometer or a digital voltmeter. When there is no current flow, there is no waste of energy. The potential that is measured is considered to be the maximum.

Although we can never find the maximum work from a cell reaction, the value obtained is still useful when evaluating the efficiency of real processes based on the cell reaction. The charge on 1 mole of electrons is a constant which is called the Faraday. It is abbreviated as F. It has the value 96,485 coulombs of charge per mole of electrons. The value q equals the number of moles of electrons times the charge per mole of electrons. In other words, the equation can be written as

$$q = n \times F$$

The efficiency of the cell is

$$\text{Efficiency} = (w/w_{max}) \times 100\%$$

Let us now relate the potential of a galvanic cell to the free energy. For processes that are carried out at constant temperature and pressure, the change in free energy equals the maximum amount of work for a process. This can be expressed as

$$w_{max} = \Delta G$$

For a galvanic cell

$$w_{max} = -q\epsilon_{max} = \Delta G$$

Since

$$q = n \times F$$

$$\Delta G = -q \times \epsilon_{max}$$

$$\Delta G = -nF \times \epsilon_{max}$$

The maximum cell potential is directly proportional to the free energy difference between reactants and products in the cell. This equation provides an experimental means in order to obtain ΔG for a reaction. It also confirms that galvanic cells will run in the direction that gives a positive value for a cell standard reduction value. A positive cell standard reduction value corresponds to a negative ΔG value which is indicated by a spontaneous process.

Cell Potential involving Concentration

We will look at how cell potential depends on concentration. We can build galvanic cells where both compartments have the same components but the concentrations are different. This is called a concentration cell. The difference in concentration is the factor that makes a cell potential. Voltages are typically small. Cell potential depends on concentration which results from the dependence of free energy on concentration. Consider the equation

$$\Delta G = \Delta G^{\circ} + RT\ln(Q)$$

This is where Q is the reaction quotient. This equation is used to calculate the effect of concentration on the change in free energy. Since

$$\Delta G = -nF\in \text{ and } \Delta G° = -nF\in°$$

we can rewrite the equation as

$$-nF\in = -nF\in° + RT\ln(Q)$$

Dividing both sides of the equation by -nF gives

$$\in = \in° - (RT/nF)\ln(Q)$$

This gives the relationship between cell potential and concentrations of the cell components. This equation is called the Nernst Equation. At room temperature, the Nernst Equation is sometimes written in the form

$$\in = \in° - (0.0591/n)\log(Q)$$

The potential from the Nernst Equation is considered to be the maximum potential before current has flowed. As the cell begins to discharge, current flows from anode to cathode. The concentrations eventually change. As a result, the cell standard reduction value will also change. The cell spontaneously discharges until it reaches equilibrium. This is the point where Q = K (the equilibrium constant) and the cell standard reduction value is equal to zero. For example, a "dead" battery is obtained when the cell reaction has reached equilibrium. This is where there are no longer driving forces to push electrons through the wire. At equilibrium, the components in the two cell compartments are considered to have the same free energy. This is when $\Delta G = 0$ for the cell reaction at equilibrium concentrations. The cell does not have the ability to do work.

Ion-Selective Electrodes

Cell potentials are sensitive to both the concentrations of the reactants and products during the cell reaction. Measured potentials are used to determine the concentration of an ion. A device called a pH meter measures concentration using an observed potential. A pH meter has three main components. These include a standard electrode with a known potential, a glass electrode that tends to charge the potential, and a potentiometer. The glass electrode that changes potential depend on the concentration of the hydronium cations in solution when it is dipped.

The potentiometer measures the potential between the electrodes. The potentiometer gets converted electronically to a pH reading for the solution being tested. The glass electrode has a reference solution of dilute hydrochloric acid that is in contact with a thin glass membrane. The glass electrode that has an electric potential depends in the difference in the concentration of the hydronium cation between the reference solution and the solution where the electrode is dipped. Electrical potential therefore varies with the pH of the solution. Ion-selective electrodes are sensitive to the concentration of a particular ion. Glass electrodes are made sensitive based on the composition of the glass. Other kinds of ions could also be detected if a crystalline surface replaces the glass membrane itself. A cell at equilibrium has the following relationship when $\epsilon_{cell} = 0$ and when Q = K. Using the Nernst Equation at room temperature,

$$\epsilon = \epsilon° - (0.0591/n)\log(Q)$$

$$0 = \epsilon° - (0.0591/n)\log(K)$$

$$\log(K) = n\epsilon°/0.0591$$

Batteries

A battery is considered to be a galvanic cell or a group of galvanic cells connected in series. The potential of the individual cells add up to give the total battery potential. Batteries are known to be a source of direct current and for being portable in society. The lead storage battery has been around for quite a long time. It has been important for the automobile to be used for transportation. The battery can be used for temperature extremes from very low temperatures to high temperatures above one hundred degrees Celsius. For this type of battery, lead serves as the anode and the lead coated with lead dioxide serves as the cathode. Both of these electrodes are dipped into an electrolyte solution containing sulfuric acid. The electrode reactions are as follows

Anode Reaction: $Pb(s) + HSO_4^-(aq) \longrightarrow PbSO_4(s) + H^+(aq) + 2e^-$

Cathode Reaction: $PbO_2(s) + HSO_4^-(aq) + 3H^+(aq) + 2e^- \longrightarrow PbSO_4(s) + 2H_2O(l)$

Cell Reaction: $Pb(s) + PbSO_2(s) + 2H^+(aq) + 2HSO_4^-(aq) \longrightarrow 2PbSO_4(s) + 2H_2O(l)$

The lead storage battery has six cells that are connected in series. Each cell has multiple electrodes in the form of grids. Sulfuric acid gets consumed as the battery discharges. The density of the electrolyte solution gets lowered. We can monitor the condition of the battery by measuring the density of the sulfuric acid solution. The solid lead sulfate that is formed during the reaction adheres to the grid surfaces of the electrodes. The battery can be recharged by forcing current in the opposite direction to reverse the cell reaction. A car's battery is always continuously charged by an alternator by the engine.

An automobile that has a dead battery can be "jump-started" by connecting the battery to the battery in an automobile that is running. This is dangerous. The flow of current causes electrolysis of water inside the dead battery. This produces hydrogen and oxygen gases. Removing the jumper cables after the disabled car starts can cause the gaseous mixture to be ignited. The battery will explode ejecting corrosive sulfuric acid. We can avoid this problem by connecting the ground jumper cable to a part of the engine away from the battery. Any arc produced when the cable is disconnected will be harmless.

Some storage batteries require periodic "topping off" since the water in the electrolyte solution gets depleted by the electrolysis that also accompanies the charging process. Updated types of batteries have electrodes made of an alloy of calcium and lead since these two elements inhibit the electrolysis of water. The batteries are sealed since they don't require any water added.

Other Batteries and Cells

Calculators, digital watches, and portable CD players are powered by small and efficient batteries. The dry cell battery has a zinc inner case that is supposed to be the anode. There is a carbon rod that is in contact with a moist paste of solid manganese dioxide, solid ammonium chloride, and carbon that is considered to be the cathode. Alkaline version of dry cell batteries have potassium hydroxide or sodium hydroxide instead of ammonium chloride. Alkaline dry cell batteries tend to last longer because the zinc anode corrodes less during basic conditions rather than acidic conditions. Silver cell batteries contain a zinc anode. The cathode employs silver oxide as an oxidizing agent under basic conditions. Mercury cell batteries contain a zinc anode. The cathode employed is mercury(II) oxide as an oxidizing agent under basic conditions. The cathode is steel. Mercury cells are used in calculators. Nickel-cadmium batteries are different. The products for the half reactions stick to the electrodes. Nickel-cadmium batteries can be charged as many times as possible.

Fuel cells are galvanic cells where the reactants are continuously supplied. Let us take a look at the redox reaction of methane mixed with oxygen. The products of the reaction are carbon dioxide, water, and a release of energy. The energy from this reaction gets released as heat, and this heat is used to warm homes and for us to use machines. The energy for fuel cells, however, is used to give an electric current. Electrons flow from the reducing agent which is methane to an oxidizing agent which is oxygen through a conductor. The space shuttle uses a fuel cell based on the reaction using hydrogen and oxygen to form water.

Corrosion

Corrosion is thought of as a process to return metals back to their natural state. Their natural state are from ores for which they were obtained. The process of corrosion involves the metal being oxidized. Corroded metal loses its structure and attractive qualities. Both iron and steel are used to replace rusted metal. Metals with the exception of gold have standard reduction potentials less positive than oxygen gas. Reversing the half-reactions shows the metal being oxidized. Combine this with the reduction half-reaction for oxygen, a positive cell potential value results. Metals being oxidized by oxygen is a spontaneous process. We cannot tell based on the potential how fast the reaction is occurring. Nobel metals such as copper, gold, platinum, and silver are difficult to oxidize.

Although there is a large difference in reduction of potentials between oxygen and metals, the problem of corrosion does not prevent metals being used in air. Most metals develop a coating of thin oxide which protects itself from further oxidation. Aluminum is an example that gets easily oxidized by oxygen gas. The formation of Al_2O_3 more properly reported as $Al_2(OH)_6$ inhibits and prevents any further corrosion. The formation of this oxide is used as a good standard material. Iron is another example that forms a protective oxide coating. But it is not strong enough to protect itself from corrosion. Exposing steel to oxygen in moist air, an oxide is formed and tends to scale off and new metal surfaces corrode. Nobel metals such as copper and silver have corrosion products. Copper forms a layer of greenish copper carbonate called patina. Silver sulfide or silver-tarnish gives silver a rich appearance. Gold does not corrode in air.

Steel is considered to be the main structural material used to build bridges, buildings, and automobiles. Controlling corrosion for steel is necessary if we are going to use it. Iron corroding is an electrochemical reaction. Steel is nonhomogeneous due to its chemical composition. The surface of steel is nonuniform. There are also physical strains that leave stress points in the metal. The nonuniformity surfaces is where the iron is more easily oxidized. The areas that are oxidized are the anodic regions. The other regions are the cathodic regions. In the anode region, Fe gets oxidized to Fe^{2+}. The electrons are released and flow through the steel to a cathodic region. This is where they react with oxygen. The Fe^{2+} ions are formed in the anodic regions and travel to the cathodic regions by the moisture on the surface of the steel. In the cathodic region, Fe^{2+} react with oxygen to make rust. Rust is hydrated iron(III) oxide of different compositions.

Iron also undergoes corrosion. Because ions and electrons migrate, rust forms at sites that are not in the same location where the iron dissolved to form pits in the steel. The amount of hydration of the iron oxide affects the rust color. It can vary from black to reddish-brown to yellow. Moisture is important in the corrosion process. Moisture is needed to act as a salt bridge between anodic and cathodic regions. Keep in mind steel does not rust when the air is dry. This explains why cars last longer in the arid southwest than in humid midwest. Salt is used to melt both ice and snow. It also speeds up the rusting process. This is a problem for the colder regions in the world. Rust increases because the dissolved salt on the moist steel surface increases the conductivity of an aqueous solution and accelerates the electrochemical corrosion process. Stable complexes of Fe^{3+} and Cl^- tend to get iron dissolved which is a factor of the corrosion process.

Preventing corrosion helps conserves our natural resources of energy and metals. Applying a coating such as paint or metal plating is used to protect the metal from oxygen and moisture. The elements chromium and tin are used to plate steel since they oxidize to form an effective oxide coating. Zinc is also used to coat steel by a process called galvanizing. Zinc forms a mixed oxide-carbonate coating. Zinc is more of an active metal than iron. Oxidation that takes place dissolves zinc instead of iron. Zinc acts as a coating on steel.

Forming alloys is also used to prevent corrosion. Stainless steel contain both nickel and chromium. Both form oxide coatings. Surface alloys are now being developed. A cheap carbon steel gets treated by ion bombardment to make a thin layer of stainless steel or an alloy on the surface. A plasma or a type of ion gas of the alloying ions get formed at high temperatures. It is then directed onto the surface of the metal. Cathodic protection protects steel in buried fuel tanks and pipelines. Magnesium is considered to be an active metal. It is connected by a wire to either the pipeline or tank to be protected. Magnesium is a better reducing agent than iron. Electrons are supplied by the magnesium instead of the iron. This keeps the iron from being oxidized. When oxidation takes place, the magnesium anode dissolves, and it has to be replaced periodically.

<u>Electrolysis</u>

Galvanic cells produce current when an oxidation-reduction reaction occurs spontaneously. An electric cell uses electrical energy to make chemical changes. Electrolysis involves forcing current through a cell to make chemical changes in which the cell potential is negative. Electrical work causes nonspontaneous chemical reactions to occur. Electrolysis is used for changing a battery, making aluminum metal, and chrome plating. External power source is connected forcing electrons to the cell going in the opposite direction. Both anodes and cathodes become reversed. The flow of ions in the salt bridge is opposite in the two cells. Chemical changes occur with the flow of a given current during a time period. Other important units include an ampere which is abbreviated as A. It is one coulomb of charge per second. When we talk about plating, we are depositing a neutral metal on the electrode by reducing the metal ions in solution. Reductions occur at the cathode for the electrolytic cell. Solving stoichiometry problems requires us to multiply the current and time to obtain the charge. Converting the charge to moles of electrons using Faraday's constant, we are able to convert moles of electrons to moles of the desired element and then finally converting moles to grams of the desired element.

We learned earlier that hydrogen and oxygen combine spontaneously to make water giving a decrease in free energy used to run fuel cells which then produces electricity. Reversing the process is nonspontaneous. Suppose solutions in electrolytic cells contain cations. Metals can be plated out onto the cathode if the voltage is turned up. The more positive the standard reduction value, the better the reaction has a tendency to proceed in the desired direction. Electrolysis of an aqueous solution of a salt gives the sodium cation, chloride anion, and water. The chloride anion and water are readily oxidized. Experimentally, even though the chloride anion may have a smaller reduction potential than water, it is the chloride anion that is first oxidized. Higher potential is necessary to oxidize water. Excess voltage greater than the expected value is called overcharge. Chlorine is eventually produced first. There is some difficulty transferring electrons from the species in solution to atoms on the electrode across the electrode-solution interface. We need to be careful using standard reduction potentials when we are predicting the order of redox species for electrolytic cells.

Metals have their ability to donate electrons to make ions. They are considered to be good reducing agents. Most metals are found in nature as ores, mixtures of ionic compounds that contain oxide, sulfide, and silicate anions. Nobel metals include platinum, silver, and gold. They are difficult to oxidize. They are normally found as being pure. Aluminum, for example, is one of our abundant elements on Earth. Aluminum is found in the ore called Bauxite. In 1886, Charles M. Hall in the United States and Paul Heroult in France came up with a way to produce aluminum using an electrolytic process. Molten cryolite, Na_3AlF_6, is used as the solvent for aluminum oxide. Ions have to move to the electrodes for electrolysis to take place. Ion mobility is obtained by dissolving the substance in water. But this cannot happen for aluminum since water gets more easily reduced than Al^{3+}. We could melt the salt alumina (Al_2O_3) to obtain ion mobility. But alumina melts at a temperature over two thousand degrees celsius. If alumina molten cryolite is mixed together, the melting point becomes one thousand degrees celsius. The resulting molten mixture is used to obtain aluminum metal electrolytically.

The Bauxite ore does not contain pure aluminum oxide. It also contains oxides of titanium, silicon, and iron and silicate materials. In order to obtain pure hydrated alumina ($Al_2O_3*nH_2O$), the crude bauxite gets treated with aqueous sodium hydroxide. Alumina is amphoteric. Carbon dioxide is then added to acidify the aluminate ion separated as a sludge. Hydrated alumina is then reprecipitated. The purified alumina is mixed with cryolite and then melted. The aluminum ion is reduced to the aluminum metal using electrolysis. Aluminum can then be alloyed with other metals like manganese and zinc.

Purification

Purifying metals is another application of electrolysis. Metal plating helps protect metals from being corroded by applying a thin coating of the metal. The metal resists corrosion. Sodium can be produced by electrolyzing molten sodium chloride. Mixing sodium chloride and calcium chloride lowers its melting point. The mixture can then be electrolyzed in a Downs Cell. In the Down cell reaction, sodium is a liquid and gets drained off. It is then cooled and made into blocks. Sodium is reactive so we store it in an inert solvent like mineral oil to prevent oxidation. Brine is obtained by the electrolytic process of sodium chloride. We get chlorine and sodium hydroxide. Electrolysis of brine produces both hydrogen and chlorine.

Dissolved sodium hydroxide and sodium chloride is left in the solution. Sodium hydroxide gets contaminated with sodium chloride. This can be eliminated using a mercury cell. Mercury cells can be used to electrolyze brine. Mercury acts as the conductor at the cathode. Hydrogen gas has a high over voltage with a mercury electrode, the sodium cation gets reduced instead of water. The sodium metal dissolves in the mercury. The liquid alloy gets pumped to a chamber where the dissolved sodium reacts with water to make hydrogen gas.

Pure solid sodium hydroxide is recovered from the aqueous solution. Mercury is regenerated and pumped back to the electrolytic cell. This process is called the chlor-alkali process. The process works but using mercury is dangerous. Chlor-alkali production is carried out in diaphragm cells. The diaphragm cell has the cathode and anode separated by a diaphragm where water molecules, sodium cations, and tiny bit chloride anions move. The diaphragm does not allow hydroxides anions to pass through. Hydrogen gas and hydroxide formed at the cathode are separated from the chlorine made at the anode. However, an aqueous solution of sodium hydroxide and unreacted sodium chloride is pumped from the cathode compartment. Membranes now are used to separate the anode and cathode apartments in the electrolytic cells. The membrane is impermeable to anions. Cations can flow through the membrane. Neither chloride anion or hydroxide anion can go through the membrane that separates the anode and cathode compartments. Sodium hydroxide does not get contaminated with sodium chloride at the cathode.

Conclusion

Electrochemistry has to do with changes and conversions of chemical and electrical energy through redox reactions. Electrons are transferred from a reducing agent to an oxidizing agent. Oxidation is considered to be the loss of electrons, and reduction is considered to be the gain of electrons. Redox reactions are broken into half-reactions where one of the reactions is oxidation and the other one is reduction. Electrons are transferred directly for redox reactions. No work is obtained from the chemical energy during the reaction. Separating the oxidizing agent from the reducing agent allows electrons to flow through a wire. The chemical energy can produce work.

Galvanic cells are devices for which chemical energy is transformed to electrical energy. Reversing the process where electrical energy produces chemical change is called electrolysis. This can be carried out in an electrolytic cell by forcing current through the cell. Oxidation occurs at the anode whereas reduction occurs at the cathode. The driving force is called the cell potential. Galvanic cells are carried out with positive values for the cell potential. We can assign standard reduction potentials to half-reactions by considering zero volts for the half reaction $2H^+ + 2e^- \longrightarrow H_2$.

Electrical potential is measured in volts which is defined as joules or work per coulomb of charge. Maximum work can be calculated where no current flows for the maximum cell potential. Actual work obtained is always less than the maximum work possible. Energy is wasted through frictional heating when current flows through a wire. Processes carried out at constant temperature and pressure has the free energy that equals the maximum work. Concentration cells are galvanic cells where both compartments have the same components but the concentrations are different. Electrons flow in a direction that equalizes the concentration.

The Nernst Equation relates the cell potential and the concentration of its cell components. The cell potential is equal to zero when the cell is at equilibrium. We can use this equation to find equilibrium constants for redox reactions. Batteries are galvanic cells or group of cells that are connected in series. Batteries are a source of direct current. Lead storage batteries have lead anodes, lead coated with lead(IV) oxide coated, and a solution of sulfuric acid. Dry cell batteries have a moist paste of solid manganese dioxide, carbon, and ammonium chloride as the acid version. Potassium or sodium hydroxide instead of ammonium chloride is used as the alkaline version. Zinc acts as the anode, and a carbon rod acts as the cathode. Fuels cells are also galvanic cells where the reactants are constantly being supplied. Corrosion involves metals oxidizing to form oxides or sulfides. Aluminum forms an oxide that prevents any corrosion. Corroding iron is an electrochemical process. The iron surface forms both anodic and cathodic regions. The Fe^{2+} ions formed where the anionic region migrate through the moisture to the cathodic region. Electrons also flow through the metal to cathodic regions where oxygen is reacted. We can prevent corrosion by coating iron with paint or covering it with chromium, zinc, or tin. Other prevention methods include alloying and cathodic protection. Aluminum is made by electrolyzing a molten alumina-cryolite mixture carried out by the Hall-Heroult process. Metals can be refined using electrolytic methods, and the chlor-alkali process can be used to make chlorine and sodium hydroxide.

Chapter 19
Coordination Chemistry

Magic Fluid!

Transition metals have a lot of uses in society. Iron can be used for steel. Copper can be used for electrical wiring and also for water pipes. Titanium is used for paint. The list can go on and on. Transition metals can also be used in biological processes. Iron complexes give transport and storage of oxygen. Iron and molybdenum compounds are used as catalysts for nitrogen fixation. Iron and copper are used in the respiratory cycle. Cobalt is found in Vitamin B12. Unlike the representative elements, transition elements are similar within a given period and within its group. The reason why has to do with the last electrons which are the inner electrons being added to the transition metals. These are the d electrons for the d-block and f electrons for the lanthanides and actinides. These inner electrons do not participate in bonding as well as the s and p electrons. This is why the chemistry changes very little compared to the representative elements. Their group designations has nothing to do with the transition metals behavior.

Characteristics of transition metals include having a metallic luster, high electrical, and thermal conductivities. Silver and copper are the best conductors of heat and electric current. Other transition metals include tungsten that is used for filaments in light bulbs. Mercury is used in thermometers. Titanium and iron are hard, and they are used as structural materials. Silver, copper, and gold are soft. Chromium, cobalt, and nickel form oxides. They protect the metal from further oxidation. Iron forms oxides during the corrosion process. Gold, silver, platinum, and palladium are considered to be noble metals that do not form oxides.

Other characteristics of transition metals include having more than one oxidation state. A complex ion is considered to be a transition metal cation surrounded by ligands which are molecules or ions that act as Lewis bases. Transition metal ions in the complex are colored. They absorb visible light in the electromagnetic spectrum, and many of transition metal compounds are paramagnetic. The energy of 3d orbitals for transition metal ions are very much less than that for the 4s orbital. The electrons that remain after the ion is formed occupy 3d orbitals. These 3d orbitals are lower in energy.

If we look at the first row of transition elements, they include scandium, titanium, vanadium, chromium, manganese, iron, cobalt, nickel, copper, and zinc. The maximum oxidation corresponding to their group designation occur for the first five transition metals. These include scandium, titanium, vanadium, chromium, and manganese. Going towards the right hand side of the period, the maximum oxidation does not correspond to the group designation. The oxidation state 2+ is more common for iron, cobalt, nickel, copper, and zinc. There are no higher oxidation states for these metals. These metals have their 3d orbitals lower in energy as there is an increase in nuclear charge. Electrons also get harder to remove. First ionization energy increases slowly for the first row transition metals going from left to right. But the third ionization energy involves the removal of an electron from a 3d orbital. The third ionization energy increases more gradually than the first ionization energy. This explains the decrease in energy of the 3d orbitals going across a period.

Transition metals can also act as reducing agents. Metals with the most positive potential are better reducing agents. The ability to reduce a metal decreases going from left to right for first-row transition metals. Atomic radii decreases going from left to right across the first-row transition series but not in a consistent pattern. There is an increase in radius going from the 3d to the 4d metals. However, the 4d and 5d metals are similar in size. This is due to lanthanide contraction. The lanthanide series consist of elements from lanthanum to hafnium. Electrons fill the 4f orbitals. These 4f orbitals are buried inside the atom. Adding additional electrons do not greatly affect the atomic size. Increasing the nuclear charge cause the radii of these lanthanide elements to decrease going from left to right. This contraction offsets the normal size increasing going from one principle quantum level to another one. This again explains why the 5d elements are almost similar in size to the 4d elements.

Differences between 4d and 5d elements for a group increase as we go from left to right. Applications include zirconium, which is a 4d element, and zirconium oxide resist high temperature. They can be combined with molybdenum and niobium alloys. These alloys are used for space vehicles during high temperatures in the Earth's atmosphere. Tantalum, which is a 5d element, resist attacks by body fluids. It is often used for replacing bone. Ruthenium, osmium, rhodium, iridium, palladium, and platinum are considered to be the platinum metals. They are used as catalysts for industry.

Coordination Compounds

Let us now take a look at coordination compounds. Transition metal ions form coordination compounds. They are colored and paramagnetic. A coordination compound is made up of a complex ion. A complex ion consist of a transition metal with ligands attached and sometimes counter ions to make a compound with no net charge. If counter ions are present, the compound can act as an ionic solid. Ligands are neutral molecules or ions that have a lone electron pair to form a bond to a metal ion. The ligand can act as a Lewis base and the metal ion acts as a Lewis acid. The bond that forms is called a coordinate covalent bond.

Transition metals have two types of valences. The secondary valence has to do with the ability of a metal ion binding to ligands to make complex ions. This is called the coordination number. The primary valence has to do with the ability of the metal to form an ionic bond with oppositely charged ions. This is called the oxidation state. Coordination number varies for each complex ion. The bonds formed by metal ions and ligands can vary from two to eight which depends on size, charge, and the electron configuration of the transition metal ion. Two ligands opposite of each other can form a linear complex ion. Four ligands around the central atom can form either a tetrahedral or square planar geometry with the complex ion. Six ligands around the central atom can form an octahedral complex.

Ligands that form one bond to a metal ion are referred to as a mono dentate ligand or as a unidentate ligand. Examples of mono dentate ligands include water, the cyanide anion, thiocyanate, halides, ammonia, and hydroxide. Ligands that form two bonds to a metal ion are referred to as a bidentante ligand. Examples include oxalate and ethylenediamine. Some ligands have multiple atoms with lone electron pairs that can also bond to a metal ion. These are referred to as chelating ligands or chelates. Ligands that form more than two bonds to a metal ion are referred to as polydentate. Ethylenediaminetetraacetate, which is abbreviated as EDTA, is a ligand that can form up to six bonds to a metal cation. It is considered to be a hexadentate ligand. Stable complexes can be formed with this ligand, and it is used to remove toxic heavy metals from the body. Ethylenediaminetetraacetate is found in soda, salad dressings, soaps, and in most cleaners.

Naming coordination compounds is necessary since so many are now produced and synthesized. Like ionic compounds, the cation is named right before the anion. Ligands are also named before the metal cation. When we name ligands, there is an "o" added to the root name for the anion. Prefixes such as mono-, di, tri-, tetra-, etc are used to show the number of simple ligands. Prefixes such as bis-, tris-, tetrakis-, etc are also used if the previous prefixes are already used. The central metal ion has an oxidation that is designated by a Roman numeral. If there is more than one type of ligand, the ligands are named alphabetically. Prefixes are not used for alphabetical purposes. If there is a negative charge on the complex ion, the suffix -ate becomes added to the metal's name. Latin names can also be used to identify the metal. Examples of neutral unidentate ligands include aqua, ammine, methyl amine, carbonyl, and nitrosyl. Examples of anion unidentate ligands include fluoro, chloro, bromo, dodo, hydroxo, and cyano. Examples of latin names used for metal ions in anion complexes include ferrate, cuprate, plumbate, argentate, aurate, and stannate.

Complex Ion Bonding

Valence shell electron pair repulsion model predicts the structure for simple molecules. But it does not work for complex ions. But we can state a complex ion that has a coordination number of six which has an octahedral arrangement. Coordination of four could give either a tetrahedral or a square planar arrangement. Coordination of two can give a linear arrangement. Ligands act as Lewis bases that donate a pair of electrons to the empty orbital of the metal cation that acts as a Lewis acid. A coordinate covalent bond forms between the metal and the ligand. It is the hybrid orbital from the metal cation that depends both on the number and arrangement of ligands. For linear ligands to be arranged, sp hybridization is required. A square planar ligand arrangement has dsp^2 hybridization, whereas a tetrahedral ligand arrangement has sp^3 hybridization.

Complex Ion Isomers

Two or more molecules that have the same formula but if they are arranged differently from each other, are called isomers. Isomers have the same types and numbers of atoms, but these arrangements differ from each other. They end up having different physical and chemical properties. Structural isomerism are isomers that contain the same number of atoms but the bonds differ from each other. Structural isomers can be further categorized to coordination isomers and linkage isomers. Coordination isomers involves different compositions for the complex ion. Linkage isomers have the same complex ion, but the attachment of at least one of the ligands is different. Some ligands such as the thiocyanate anion can be reversed and still be used as a ligand. Stereoisomerism have the same arrangement of bonds, but the spatial arrangements are different from each other. Stereoisomers can be further categorized into geometric or optical isomers. Geometric isomers or cis-trans isomers have their atoms or groups of atoms to be in different positions around a ring or a bond. The trans isomer has two ligands that are the same across from each other. The cis isomer has two ligands that are the same next to each other. Optical isomers are isomers that have opposite effects on plane-polarized light. Optical activity exists for molecules that are nonsuperimposable mirror images such as enantiomers. Enantiomers rotate plane-polarized light in opposite directions. They are considered as optical isomers. Isomers that rotate plane-polarized light to the right are called dextrorotatory and is abbreviated as d. Isomers that rotate the plane-polarized light to the left are called levorotatory and is abbreviated as l. Equal mixtures of d and l forms are called racemic mixture. Racemic mixtures do not rotate the plane of polarized light. The opposite directions oppose each other. Geometric isomers can either be superimposable or nonsuperimposable depending on the structure of the molecule.

Crystal Field Theory

Color and magnetism of complex ions take place because of changes in the energies of the d orbitals from the metal ion caused by metal-ligand interactions. Crystal field theory deals with the energies of these d orbitals. Ligands are assumed to be negative point charges and the metal-ligand bonding can be thought of as ionic. For octahedral complexes, the d_{z2} and d_{x2-y2} orbitals point their orbitals directly at the point-charge ligands. The d_{xy}, d_{yz}, and d_{xy} point their lobes between point charges. Negative point-charge ligands repel the negatively charged electrons. Electrons fill the d orbitals farther from the ligands in order to minimize repulsions. The d_{xy}, d_{yz}, and d_{xy} orbitals make up the t_{2g} set. They are lower in energy for the octahedral complex. The d_{z2} and d_{x2-y2} orbitals make up the e_g set. Negative point-charge ligands increase the energies of the d orbitals. The orbitals that point towards the ligands are raised in energy more than those that point between the ligands.

The splitting of 3d orbital energies is symbolized as Δ which explains both color and magnetism for complex ions for the first-row transition metal cations. There are two ways to place electrons in these 3d orbitals that are split. Large splitting produced by the ligands are considered to be strong-field. Electrons pair in the lower energy t_{2g} orbitals. This is referred to as a diamagnetic complex or a low-spin complex. The minimum number of unpaired electrons are present if they are any. Small splitting produced by the ligands are considered to be weak-field. Electrons occupy all five orbitals before pairing takes place. The unpairing of electrons is referred to as a paramagnetic complex or as a high-spin complex. The maximum number of unpaired electrons is present.

For octahedral complexes, ligands can be arranged by order of their ability to produce d-orbital splitting. The spectrochemical series include the following from strong-field ligands to weak field ligands

$$CN^- > NO_2^- > en > NH_3 > H_2O > OH^- > F^- > Cl^- > Br^- > I^-$$

These ligands are arranged based on decreasing splitting of the 3d orbital energies for a given metal ion. The magnitude of the splitting of the 3d orbital energies for a given ligand increases as the metal ion's charge increases. As the charge on the metal ion increases, ligands are drawn closer to the metal cation due to the charge intensity being increased. Ligands moving closer cause greater splitting of d orbitals. A larger splitting of the 3d orbital energies is produced. These explanations account for the magnetic properties for octahedral complexes.

Let us take a look at the color explanations for complex ions. When a substance absorbs specific wavelengths of light in the visible region of the electromagnetic spectrum, the color for the substance is determined by wavelengths of visible light that remain. A substance exhibits the complementary color that is absorbed. A complex ion absorbs the specific wavelength of visible light which has to do with the transfer of a lone d electron between the split d orbitals. A molecule absorbs a photon of light if the wavelength of the light is the exact energy that is needed. The splitting of the d orbitals for most octahedral complexes corresponds to energies of photons in the visible region. Most octahedral complexes are colored. Ligands next to the metal cation determine the size of the splitting of the d-orbtials. Color change as ligands change. This happens because changing the splitting of the 3d orbital energies means the wavelength of light changes for electron transfer to occur between the t_{2g} and e_g orbitals.

Let us take a look at tetrahedral arrangements of point charges for 3d orbitals of a metal cation. None of the 3d orbitals are directed at the ligands in a tetrahedral arrangement. Difference in energy for the splitting of the d orbitals is a lot less for tetrahedral complexes. The tetrahedral splitting is 4/9 of the octahedral splitting for a ligand and the metal cation. We can represent this in equation form as

$$\Delta_{tet} = (4/9)\Delta_{oct}$$

The d_{xy}, d_{xz}, and d_{yz} orbitals do not exactly point to the ligands. These three orbitals are closer to point charges than the other two orbitals namely d_{z2} and d_{x2-y2}. Tetrahedral d-orbital splitting is opposite to the octahedral arrangement. The d-orbital splitting is small for tetrahedral. The high-spin complex or weak-field complex applies to the tetrahedral arrangement. There are no ligands strong enough to make a low-spin complex or strong-field complex.

For the square planar complex, the d_{x2-y2} orbital is higher in energy because it is this orbital that points at the four remaining ligands. The d_{xy} orbital is higher than the d_{z2} orbital. Below the d_{z2} orbital are a degenerate pair of the orbitals d_{xz} and d_{yz}. A linear complex is arranged by having two ligands along the z-axis and removing the other four in the xy plane. The d_{z2} orbital ends up pointing at the ligands, and it is considered to be higher in energy than the other d orbitals. The next pair of d-orbitals are a degenerate pair of d_{xz} and d_{yz}. The lowest pair of degenerate d-orbitals are d_{xy} and d_{x2-y2}.

Conclusion

First-row transition metals end up having their last electrons fill in the 3d orbitals. Electrons in these inner orbitals do not get involve in bonding as well as the s and p orbitals. The chemistry that takes place for transition metals are not that much affected by adding electrons to each of its elements. Transition metals have more than one oxidation state. Cations become complex ions. Most of these compounds are colored and are found to be paramagnetic. A variety of cations can be formed by losing one or more electrons. Maximum possible oxidation state results in losing the 4d and 3d electrons. However, this does not hold true towards the right hand side of the first row transition metals. A positive charge of 2+ is more common because the 3d electrons are difficult to remove.

Transition metals can form coordination compounds. These consist of a complex ion, which includes a transition metal with ligands attached, and counter ions that leaves the solid with no overall charge. Ligands are Lewis bases that can be neutral and have a lone pair of electrons which shares with an empty p-orbital of a metal cation and which acts as a Lewis acid to form a coordinate covalent bond. Coordination number is defined as the number of bonds between the metal and the ligands. This depends on size, charge, and electron configuration for the metal cation. Octahedral arrangements have a coordination number of 6, linear arrangements have a coordination number of 2, and a tetrahedral arrangement or a planar arrangement has a coordination of 4. Ligands form different types of bonds. Unidentate ligands form one bond. Bidentate ligands form two bonds. Hexadentate ligands form six bonds. Chelates are ligands that bond to the metal cation through more than one atom.

Two species or more can have the same formula, but they can have different physical and chemical properties. These are called isomers. Structural isomers have the same atoms, but the bonding arrangement is different. An example of structural isomerism is coordination isomerism. The coordination around the metal cation is different. Linkage isomerism have the same coordination, but the ligands can be attached differently. Stereoisomerism has to do with isomers having the same bonds, but the spatial arrangements are different. Geometric isomers have atoms or groups of atoms in different positions around the metal atom. Identical ligands that are found across from each other are referred to as trans. Identical ligands that are found next to each other are referred to as cis. Optical isomerism has to do with nonsuperimposable mirror images. Enantiomers can be formed which rotate plane-polarized light but the directions are opposite.

Bonding for complex ions can be described by the localized electron model based on hybridization. But the model fails to account for properties such as magnetism and color. Crystal field theory accounts for both magnetism and color. Energies of the d orbitals of the metal cation are involved. Ligands have effects on the energies of the d orbitals. Ligands can be assumed as negative point charges. Metal-ligand bonding is considered to be ionic. Point charges on the octahedral complex ion split the 3d orbitals into two different groups with different energies. Large splittings give strong-field complexes or low-spin complexes. Small splittings give weak-field complexes or high-field complexes. Ligands are arranged based on how they split the d orbitals according to the spectrochemical series. Color is obtained for complex ions based on absorbing a specific wavelength of light in the visible region. The color is determined by wavelengths of visible light that are not absorbed. Complex ions absorb light. As they absorb light, electrons are transferred between the d orbitals that are split.

Chapter 20

Organic Compounds

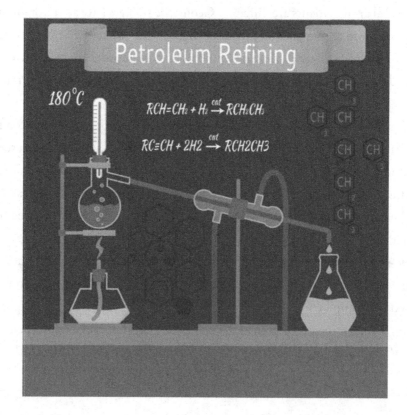

Educational Chemistry Laboratory: Installation for the Distillation of Crude Oil

Hydrocarbons

Carbon is one of the few elements that form most natural substances. Carbon has the ability to bond to itself and form long chains or rings. This element can also form covalent bonds to other nonmetals such as hydrogen, oxygen, nitrogen, sulfur, and halogens. Organic chemistry is the study of carbon-containing compounds. Synthetic fibers, plastics, artificial sweeteners, and medicinal drugs all contain carbon atoms linked to each other. Energy is based on organic materials found in coal and petroleum. Hydrocarbons are made up of carbon and hydrogen. Single carbon-carbon bonds are saturated. All of the carbon atoms are bonded to four atoms. This is the maximum number it can bond to. Hydrocarbons that have multiple bonds are unsaturated. Unsaturated hydrocarbons are more reactive than saturated hydrocarbons. Hydrocarbons can also form rings to make cyclic compounds. Unsaturated hydrocarbons that has at least one carbon-carbon double bond are called alkanes. Unsaturated hydrocarbons that have at least one triple carbon-carbon bond are called alkynes.

Since alkenes and alkynes are unsaturated, they undergo addition reactions. Hydrogenation reactions involve adding hydrogen atoms to an alkene to form a saturated compound. Halogenation involves the addition of hydrogen atoms to unsaturated compounds to form saturated compounds. Aromatic hydrocarbons are both cyclic and unsaturated. Examples of complex aromatic compounds include naphthalene which is used in mothballs, anthracene which is used for dyes, phenanthrene which is used in dyes and the synthesis of drugs, and 3,4-Benzopyrene which is an active carcinogen found in both smoke and smog.

Functional Groups

Most organic compounds contain the elements carbon, hydrogen, and another element. These are considered to be hydrocarbon derivatives. These hydrocarbons that have additional atoms are called functional groups. Examples of other functional groups include halides where a carbon atom is bonded to a halogen, an alcohol where a carbon atom is bonded to a hydroxyl group, an amine where a carbon atom is bonded to an amino group, a nitrile where a carbon atom is bonded to a nitrile group, a nitro where a carbon atom is bonded to a nitro group, and a thiol group where a carbon atom is bonded to a thiol group. Sulfides are formed where a sulfur atom is bonded to two carbon atoms. A sulfide has two carbons and in addition an oxygen atom is bonded to sulfur. Sulfones have sulfur bonded to two carbons and two oxygen atoms.

The Carbonyl Group

Of all functional groups, the carbonyl group is more common. A carbonyl group contains a carbon atom double bonded to an oxygen atom. Aldehydes contain a carbonyl bonded to a carbon atom and to a hydrogen atom. Ketones contain a carbonyl carbon atom bonded to two carbon atoms. A carboxylic acid contains a carbonyl bonded to a carbon atom and a hydroxyl group. Esters contain a carbonyl bonded to a carbon atom and an oxygen atom that is also bonded to a different carbon atom. Amides consist of a carbonyl bonded to carbon and an amino group. Carboxylic acid chlorides consist of a carbonyl bonded to a carbon atom and to a chlorine atom. Carboxylic acid anhydride consist of a carbon atom bonded to a carbonyl. The carbonyl is then bonded to oxygen and is then bonded to another carbonyl group. The last carbonyl group is then bonded to a carbon atom.

Haloalkanes and Oil

Let us take a look at the uses of haloalkanes. Haloalkanes are used as solvents and anesthetics. Dry cleaners use dichloromethane, 1,1,1-trichloroethane, and 1,1,2-trichloro-1,2,2-trifluoroethane as solvents. Examples of anesthetics include fluothane and chloroethane. Chloroethane is applied to the skin and then evaporates rapidly. The skin feels cooled and sensation is lost. Anesthetics are also used to be inhaled or injected to the patient. This causes a loss of sensation. This is done so that when surgery or other medical procedures are carried out, the patient won't feel any pain. Anesthetics are nonpolar compounds so they are soluble in nerve membranes. They decrease the ability of nerve cells to feel pain to the patient. Other anesthetics include forane, ethrane, and enthrone which have been developed that does not cause nausea.

Another application of alkanes include the use of crude oil. Crude oil or petroleum is made up of a lot of hydrocarbons. Oil refinery is a place where the components of crude oil get separated by fractional distillation. Fractions of hydrocarbons are separated by heating the mixture to high temperatures. Larger hydrocarbons require higher temperatures to heat to there boiling point before they become gases. Boiling one to four carbon atoms gives natural gas. Boiling five to twelve carbons gives gasoline. Boiling twelve to sixteen carbon atoms gives kerosene and jet fuel. Boiling fifteen to eighteen carbon atoms gives diesel fuel and heating oil. Boiling eighteen to twenty-five carbons gives lubricating oil. Boiling over twenty-five carbons gives asphalt and tar. Gases get removed and pass through a distillation column. The mixture cools and condenses to the liquid state.

Characteristics of Compounds

Specific functional groups attached to alkanes gives them different chemical and physical properties. Alcoholic beverages contain ethyl alcohol. Isopropyl alcohol is used to disinfect the skin before treating cuts. Acetone is used as fingernail polish remover. Butyraldehyde has a butter taste to many foods and margarine. Acetic acid has the functional group carboxylic acid that makes up vinegar. Aspirin, Acetaminohen, and Ibuprofen contain the functional groups carbonyl and an aromatic ring. Fruits contain the functional group esters. Esters are also found in household cleaners, glues, and polishes. Methyl amine is found in fish. Alkaloids such as caffeine, nicotine, histamine, and epinephrine are made by plants to keep insects and animals away. Odors such as lemon, orange, lavender, and rose are due to volatile compounds. These volatile compounds are made by plants. Unsaturated compounds give off the pleasant flavors and fragrances of fruits and flowers. Limonene and myrcene give the odors and flavors to lemons, oranges, and bay leaves. Citronellal gives the odor of lemon grass. Geraniol gives the odor of roses. These compounds are extracted or synthesized and are used as perfumes and flavorings.

Alkenes are also used in insect communication. Insects and other organisms emit pheromones. Pheromones are used by insects to send messages to their same species. Phermones are used to warn danger, make a trail, or attract the opposite sex. Bombykol is an example of a pheromone. It is a sex pheromone made by the female silkworm moth species. It is sixteen carbon atoms long with a cis double bond, a trans double bond, and an alcohol. The configuration of the isomer is important for it to work as a sex attractant for the silkworm moth. One change in configuration of the isomer will result in the attractant not to work. Alkenes are not only found in the insect world. Alkenes are also found in the rod and cone cells in the retina of our eyes. Rods on the edge of the retina help us see in dim light. Cones are in the center which help us see in bright light. The rods have a substance called rhodopsin which absorbs light. Rhodopsin is made up of cis-11-retinal which is attached to a protein. Rhodopsin absorbs light to change cis-11-retinal to 11-trans-retinal. The shape changes. The trans form of the isomer does not fit in the protein. The isomer 11-trans-retinal separates from the protein. An electrical signal is generated where the brain converts it to an image. The enzyme isomerase converts the trans isomer back to the cis isomer and rhodopsin reforms back to its original shape. Any lack of rhodopsin in the rods of the retina can cause us not to see in the dark. We need Vitamin A from Beta-Carotene. We get Beta-Carotene from carrots, squash, and spinach. Beta-Carotene gets converted to Vitamin A in the small intestine which is either converted to cis-11-retinal or stored in the liver.

Alkenes are also used for hydrogenation. Examples of vegetable oils are corn oil and safflower oil. These compounds are unsaturated fats made up of fatty acids that have double bonds. Hydrogenation involves converting the double bonds in these unsaturated fats to saturated fats such as margarine. Sometimes partially hydrogenated fats can be produced to make soft margarine and shortenings for cooking. Oleic acid is found in olive oil and has a cis-double bond at carbon 9. Hydrogenating oleic acid produces stearic acid which is a saturated fatty acid.

Oxygen and Sulfur Compounds

Organic compounds also consist of oxygen and sulfur. Methanol is found in solvents and paint removers. It is used to make plastics, medicines, and fuels. Ethanol is formed from grains and starches by fermentation. Ethanol is used as a solvent for perfumes, varnishes, and medicines. Oils of plants are derivatives of phenol. Eugenol is in cloves, vanillin is in vanilla bean for flavoring, isoeugenol is in nutmeg, and thymol is found in mint. Thymol has a minty taste and is found in mouthwashes. Dentists use it to disinfect a cavity before a filling is added. Glycerin is a trihydroxy alcohol. It is viscous that is obtained from oils and fats during the making of soaps. The alcohol groups on glycerin make it polar and soluble in water. There are two primary alcohols and one secondary alcohol group. This is why it is used as a skin softener in skin lotions, liquid soaps, shaving creams, and cosmetics. Ethylene glycol also contains alcohol groups. Specifically, there are two primary alcohol groups. It is used as an antifreeze for heating and cooling systems. It is a solvent for paints, plastics, and inks.

Organic compounds also include aldehydes, ketones, and carboxylic acids. Benzaldehyde is found in almonds, vanillin is found in vanilla, cinnamaldehyde is found in cinnamon, muscle is found in musk perfumes, and spearmint oil is found in carvone. Alpha hydroxy acids are carboxylic acids found in fruit, sugarcane, and milk. They are used to remove acne scars, reduce irregular pigmentation, and reduce age spots. Alpha hydroxy acids are added to skin care products for smoothing fine lines, to improve texture, and to clean pores. Examples of alpha hydroxy acids include glycol acid, lactic acid, tartaric acid, malice acid, and citric acid. Glycolic acid is found in sugarcane and sugar beet. Lactic acid is found in sour milk, tartaric acid is found in grapes, malic acid is found in apples and grapes, and citric acid is found in citric fruits such as lemons, oranges, and grapefruits.

Products that have alpha hydroxy acids increase the sensitivity of the skin from the sun and ultraviolet radiation. Sunscreen with a sun protection factor (SPF) of at least 15 is used when treating the skin with products that have alpha hydroxy acids. The concentration of the alpha hydroxy acid should be less than 10% and the pH value should be greater than 3.5. The Food and Drug Administration (FDA) feels that alpha hydroxy acids can cause skin irritation such as blisters, rashes, and discoloring of the skin. The FDA advises us to test products that contain alpha hydroxy acids on a small surface area on the skin before using it on a larger surface area of the skin.

The functional group carboxylic acid is found in salicylic acid. The functional group ester is found in acetylsalicylic acid which is also known as aspirin. Aspirin can be used as an analgesic, antipyretic, and anti-inflammatory agent. Oil of wintergreen contains methyl salicylate. It has a spearmint odor and flavor. It is used in skin ointments because it passes through the skin. It is used as a counterirritant to produce heat in order to soothe sore muscles. Soap also contains carboxylic acids. Fatty acids are long chain carboxylic acids that undergo saponification with a strong base such as sodium hydroxide in lye. Coconut oil can also be used as a fat to make soaps. Perfumes are also added to soap to give a nice scent. Soap is basically the salt of a long-chain fatty acid. Both ends of the soap have different polarities. The long hydrocarbon chain is nonpolar and is soluble in nonpolar solvents like grease or oil. The carboxylate salt is ionic, and it is also hydrophilic. It is polar and soluble in water. Soap cleans oil or grease by having the nonpolar ends dissolve in nonpolar fats and oils that have dirt. The salt or ionic end dissolves in water. Soap molecules surround the grease or oil to form clusters called micelles. The salt ends are ionic and that is why they are soluble in water. Fat and oil that is coated with soap molecules are rinsed away.

Amines and Amides

Amines and amides are also another functional groups. Histamine is an example of an amine. Our bodies increase the production of histamine due to allergic reactions or injury to our cells. Blood vessels dilate and the permeability of the cell increases. This explains why there is swelling on the skin. Diphenhydramine is an antihistamine that blocks the effects of histamine. Biogenic amines are hormones that carry messages between the central nervous system and nerve cells. Examples of biogenic amines include adrenaline and noradrenaline. They are released by the adrenal medulla to raise the blood glucose levels and move blood to the muscles. Noradrenaline contract capillaries in the mucous membranes found in the respiratory passages to counteract colds, hay fever, and asthma. If there is a deficiency of dopamine, a person could be diagnosed with Parkinson's disease. Benzedrine and Neo-Synephrine reduces respiratory congestion from colds, hay fever, and asthma.

Caffeine is another example of an amine found in plants. It is considered to be an alkaloid and stimulates the central nervous system. It is found in cocoa, chocolate, soft drinks, tea, and coffee. It makes people alert, but it could cause nervousness and insomnia. It is also used in pain relievers to counteract drowsiness from antihistamines. Quinine is used to treat malaria. Atropine is used as an anesthetic for eye examinations. Codeine is used as painkillers and is found in cough syrups. Urea is an amide that is used as a component of fertilizer. It increases nitrogen in the soil. Phenacetin and acetaminophen is used in Tylenol. Acetaminophen reduces both fever and pain, but there is very little anti-inflammatory effect.

Conclusion

Hydrocarbons consist of both carbon and hydrogen. They do not contain functional groups, and they are inert. They can be straight chained or branched. Modification of hydrocarbons consist of functional groups attached.   Functional groups are attached to the molecule and is composed of atoms or group of atoms that has a specific chemical reactivity. The carbonyl group is the most common functional group that is present in aldehydes, ketones, carboxylic acids, esters, amides, carboxylic acid chlorides, and carboxylic acid anhydride. Other functional groups include halides, alcohols, ethers, amines, nitriles, nitro, sulfides, sulfoxide, sulfone, and thiol. Applications of these kinds of functional groups include alcoholic beverages, cosmetics, foods, medicines, and insect communication.

Alkyl halides or haloalkanes are used as industrial solvents or inhaled as anesthetics in medicine. Major sources of alkanes is natural gas and petroleum. Crude oil consist of complex mixtures of hydrocarbons. Fractional distillation of crude oil gives natural gas, gasoline, kerosene, jet fuel, diesel fuel, heating oil, lubricating oil, asphalt, and tar. Flavors and odors of foods and household products is due to functional groups of organic compounds. Fragrance of organic compounds has to do with the functional group alkenes. Isomerism takes place for cis-trans retinal in rhodopsin. Hydrogenation of unsaturated fats to saturated fats takes place for alkenes. Alcohols can be used as solvents whereas a dialcohol can be used as antifreeze. Ethers can be used an anesthetics. Vanilla has been used as a flavoring for thousands of years. Aromatic aldehydes are used to give flavor for food and are used as fragrances in perfumes. Alpha hydroxy acids are carboxylic acids used for the skin. Aspirin contains an aromatic ring, a carboxylic acid, and an ester. Soaps contain polar and nonpolar regions where the polar region is the outer surface layer whereas the nonpolar region is the inner layer. Micelles are formed and are then rinsed away. Amines and amides are important for our health and for medicine. Alkaloids are amines found in plants.

Chapter 21
Earth's Chemistry

Satellite View of Planet Earth

The Atmosphere

Our atmosphere is a region above the surface where chemistry takes place. A lot of it comes from solar radiation, natural events, and activities civilization has carried out. Different physical processes occur in our atmosphere such as vertical mixing, boundary layers, inversion layers, mixing times, and transport. Even different reactions of organic compounds such as alkanes, alkenes, aromatic hydrocarbons, and hydrocarbons with different functional groups take place in the atmosphere. Two of the main gases in the atmosphere are oxygen and nitrogen. There are trace amounts of other gases too. Our atmosphere is rich in both oxygen and nitrogen. Oxygen and nitrogen cycles take place between the surface and the atmosphere. Other examples of cycles that take place in the atmosphere include the Chapman Odd-Oxygen Reactions, The OH_x Catalytic Cycle, The NO_x Catalytic Cycle, The Methane Oxidation Cycle as well as other cycles that take place.

The atmosphere is divided up into several different layers based on temperature, variation, and composition. Above land is the troposphere. The troposphere contains most of the total mass and almost all of the water's vapor. It is the thinnest layer of the atmosphere. Natural events such as rain, snow, hurricane, and lightning occur in the troposphere. Temperature decreases as altitude increases in the troposphere. The stratosphere is above the troposphere. The stratosphere consist of oxygen, nitrogen, and ozone. Air temperature rises with increasing altitude in the stratosphere. The mesosphere is above the stratosphere. Temperature decreases with increasing altitude in the mesosphere. The ionosphere or thermosphere is the top layer of the atmosphere.

Effects of the Atmosphere

Ozone is being depleted in the atmosphere. Ozone that is present in the stratosphere prevents ultraviolet radiation from reaching the surface of the Earth. Oxygen undergoes photodissociation by solar radiation. Reactive oxygen atoms combine with oxygen gas to form ozone. An inert gas such as nitrogen is to absorb some excess energy released and prevent ozone undergoing spontaneous decomposition. Any excess energy that is not absorbed by an inert gas gets released as heat. Ozone absorbs ultraviolet radiation to form oxygen and oxygen atoms. The process cycles back again by combining an oxygen atom and an oxygen molecule to make ozone. The stratosphere ends up getting warm. Ozone also protects us against ultraviolet radiation which could cause skin cancer and destroy crops on Earth. We can represent the reactions as the following

$$O_2 + hv \longrightarrow O + O$$

$$O + O_2 \longrightarrow O_3$$

$$O_3 + hv \longrightarrow O + O_2$$

Chlorofluorocarbons, which are also called Freons, has an effect on the ozone layer. Examples include Freon 11 which is $CFCl_3$ and Freon 12 which is CF_2Cl_2. They have been used as coolants in refrigerators, air conditioners, solvent cleaning, aerosols, and foam insulation. These chlorofluorocarbons are inert and diffuse slowly to the stratosphere. It takes years for these chlorofluorocarbons to reach the stratosphere. Ultraviolet radiation of chlorofluorocarbons in the stratosphere cause them to decompose. Reactive chlorine atoms can eliminate close to one hundred thousand ozone molecules. The reactions can be represented as the following

$$CFCl_3 \longrightarrow CFCl_2 + Cl$$

$$CF_2Cl_2 \longrightarrow CF_2Cl + Cl$$

$$Cl + O_3 \longrightarrow ClO + O_2$$

$$ClO + O \longrightarrow Cl + O_2$$

Overall Result: $O_3 + O \longrightarrow 2O_2$

Nitrogen oxides can also destroy stratospheric ozone. Nitrogen oxides come from exhausts of aircraft and activities that take place on Earth. Other nitrogen oxides get decomposed to nitric oxide by solar radiation. Nitrogen dioxide can also react with chlorine monoxide to make chlorine nitrate. Chlorine nitrate also gets depleted in the stratospheric ozone layer. The reactions can be represented as the following

$$O_3 \longrightarrow O_2 + O$$

$$NO + O_3 \longrightarrow NO_2 + O_2$$

$$NO_2 + O \longrightarrow NO + O_2$$

Overall Result: $2O_3 \longrightarrow 3O_2$

$$ClO + NO_2 \longrightarrow ClONO_2$$

Polar vortex is a stream of air that circles Antartica during the winter. Air that gets trapped in the vortex becomes very cold at polar night. Ice particles known as stratospheric colds form. These polar stratospheric clouds act as heterogeneous catalysts for reactions that convert hydrochloric acid and chlorine nitrate to more chlorine molecules. The reaction can be represented as the following

$$HCl + ClONO_2 \longrightarrow Cl_2 + HNO_3$$

During spring, sunlight splits chlorine to their atoms as in the following reaction

$$Cl_2 + hv \longrightarrow 2Cl$$

Ozone levels have declined by a small percent in this past decade. Dust-sized particles and sulfuric acid aerosols also enter into the atmosphere. The particles have the same catalytic function as ice crystals in the South Pole. Due to these events, the Arctic hole will grow larger as years progress. Finding other chlorofluorocarbons is still being investigated.

Hydrochlorofluorocarbon-123, which is represented by CF_3CHCl_2, is susceptible to oxidation in the lower atmosphere. It never reaches the stratosphere. The hydroxyl radical reacts with this compound to form a CF_3CCl_2 fragment. This fragment reacts with oxygen to decompose into carbon dioxide, water, and the hydrogen halides which then gets removed by rainwater. Hydrofluorocarbons can also be used to replace chlorofluorocarbons. These molecules do not contain chlorine. If they enter into the stratosphere, they won't disrupt the ozone layer. Examples of hydrofluorocarbons include CF_3CFH_2 and CF_3CF_2H.

We would like to reduce the amount of chlorine atoms in the atmosphere that is causing the depletion of the ozone layer. Having airplanes spray huge tons of hydrocarbons over the South Pole could heal the ozone layer. A chlorine atom reacting with a hydrocarbon can form hydrochloric acid and another hydrocarbon. The chlorine atom is therefore removed from the atmosphere. These products do not affect the amount of ozone present. Making large quantities of ozone and releasing it from airplanes might also be a possibility to heal the ozone layer. Even active volcanoes are a problem reducing the ozone layer. About two-thirds of sulfur found in air comes from volcanoes. Sulfur dioxide that is formed in the atmosphere becomes oxidized to sulfur trioxide which is then converted to sulfuric acid aerosols. Ozone is depleted in the stratosphere and affects climate. Aerosol clouds absorb solar radiation and cause temperature to drop on Earth's surface.

The Greenhouse Effect

Carbon dioxide present in our atmosphere also controls our climate. The trapping of heat at the Earth's surface by gases such as carbon dioxide is referred to as the greenhouse effect. The transfer of carbon dioxide from land to the atmosphere is part of the carbon cycle. Carbon dioxide is produced from hydrocarbons reacting with oxygen. Carbonate compounds can also produce carbon dioxide when heated or when it reacts with acid. Carbon dioxide can also be formed from the fermentation of sugar and as the product in respiration. It can be removed from the atmosphere by photosynthesis. Carbon left over after plants and animals die can be oxidized to carbon dioxide too. Equilibrium also exists between carbon dioxide and carbonates in both lakes and oceans. Concentration of carbon dioxide has been rising due to the burning of fossil fuels such as coal, natural gas, and petroleum. Carbon dioxide, water, methane, nitrogen oxide, nitrogen dioxide, dinitrogen oxide, and chlorofluorocarbons have also contributed to the warming of the atmosphere.

More nations that become industrialized result in increased production of carbon dioxide. If we can improve energy efficiency from automobiles, houses, and use photovoltaic cells, we can reduce the amount of greenhouse gases in our atmosphere. Recovering methane gas and reducing leakage of natural gas could help control carbon dioxide production. Protecting our forests is important in order to maintain carbon dioxide levels. Getting rid of the forest could change climate patterns on Earth. Other problems we encounter in our atmosphere is acid rain. Sulfur dioxide comes from the roasting of metal sulfides. Sulfur dioxide reacts with a hydroxyl radical to form a combined radical. The reaction can be represented as the following

$$OH + SO_2 \longrightarrow HOSO_2$$

This combined radical gets further oxidized to sulfur trioxide which reacts rapidly with water to form sulfur acid. The reactions can be represented as the following

$$HOSO_2 + O_2 \longrightarrow HO_2 + SO_3$$

$$SO_3 + H_2O \longrightarrow H_2SO_4$$

Acid rain can corrode both marble and limestone which is calcium carbonate. Even sulfur dioxide can attack marble and limestone directly. Removing sulfur from fossil fuels before combustion takes place could solve the problem, but the process is not easy. The best way is to put powdered limestone mixed with coal in a furnace to decompose calcium oxide. Calcium oxide reacts with sulfur dioxide to form calcium sulfite which is left in the purification chamber right before gases escape through the smokestack.

Photochemical smog is another problem in our atmosphere. Smog is defined as the combination of smoke and fog. The cause of smog was sulfur dioxide. Photochemical smog comes from the exhaust of automobiles when there is sunlight. Exhaust from automobiles are made up of carbon monoxide, nitrogen monoxide, and hydrocarbons. These are the primary pollutants that get converted to secondary pollutants. Secondary pollutants include nitrogen dioxide and ozone that builds up smog. Automobiles are now equipped with catalytic converters to oxidize carbon monoxide and hydrocarbons to water vapor and carbon dioxide. These converters can also reduce nitrogen monoxide and nitrogen dioxide to nitrogen gas and oxygen gas. Coating automobile radiators and air conditioner compressors with a transition metal such as platinum can help eliminate smog. Ozone and carbon monoxide as well as the use of platinum as a catalyst converts to oxygen and carbon dioxide.

Indoor pollutants such as radon is another problem. Radon is an intermediate product from the radioactive decay of uranium-238. Radon comes mostly from phosphate minerals of uranium. Radon-222 is the most dangerous. It emits alpha particles and decays into radioactive polonium. These radioactive particles can attach to smoke and dust. If they are inhaled, they can cause lung cancer. The best way to get rid of radon is to install a ventilation duct to make sure air goes from beneath the basement floor to the outside. Carbon monoxide and carbon dioxide are also another source of problem. If there is too much carbon dioxide in a sealed environment, a person can become fatigued and not concentrate very well. Proper ventilation is needed if there is too much carbon dioxide. Carbon monoxide is very poisonous. It can bind to hemoglobin in our blood. If it is tightly bound to hemoglobin, hemoglobin cannot carry oxygen for metabolism. Drowsiness, headache, and even death can occur. The best solution is to remove the person from the environment and put the victim in a place where there is plenty of oxygen and administer cardiopulmonic resuscitation.

Composition of Water

Chemical compositions of natural waters depend on chemical, physical, and biological processes. The physical and chemical changes of rocks by groundwater, the exchange of gases in the atmosphere, chemical reactions in sediments, and in mixing processes. Rocks can be broken down into smaller pieces by physical weathering. The causes of physical weathering include wind abrasion or the roots of plants that penetrate rocks. Fragments can be carried away by wind and water. They can enter into oceans and lakes. Rocks and soil can also react with acids and oxygen in water due to chemical weathering. These dissolved ions can be transported to oceans and lakes.

Gases such as oxygen, carbon dioxide, and atmospheric pollutants can be exchanged between water and the atmosphere. Exchange processes can increase with wind speed. The concentration and its solubility are also factors that are taken into account. Particles can be exchanged between water and land. Acid-base reactions, adsorption reactions, complexation reactions, dissolution-precipitation reactions, and redox reactions are reactions that can take place. Water cycles through the atmosphere and with aquatic systems. This is important for the transport and distribution of chemicals in the environment. The hydrologic cycle is the global cycle of water. Water is the most abundant liquid on Earth. Its circulation is the largest of any surface movement. Both evaporation and precipitation take place between land, ocean, and the atmosphere. Water exists in three states, and these states include gas, liquid, and solid.

Oceans and Lakes

Sunlight hits the Earth's surface and warms the surrounding air. Air density decreases and causes air to rise. A pattern of circulation is formed and mixes with the atmosphere vertically. Atmospheric winds blow east to west and vice versa. Winds drive surface currents across oceans which encounter land that causes them to move along continental borders. The spinning of the Earth causes the Coriolis force to create circulating gyres at the surface of the ocean. Oceans and lakes are heated from above. Warming decreases surface water density. The water column is stabilized. This warming effect also reduces exchange processes between warm surface waters and cold waters. Temperature decreases with depth of the ocean and then stays constant at the ocean floor. The thermocline is the region of decreasing temperature. The region of increasing density is called the pycnocline. Oceans contain surface waters and deep waters. The surface and deep regions of lakes are referred to as epilimnion and hypolimnion. They are separated by a thermocline.

Concentration of dissolved solids is higher in marine waters than fresh waters. Species that make up the dissolved solids include naturally occurring elements. The average composition of river water is a good indication of the average composition of lakes. Rivers carry chemicals in and out of lakes.

Composition of lakes is more varied than oceans. Lakes are smaller in size and are more affected by nearby atmospheric inputs and sediments. Lakes have higher surface area per volume which is more shallower in depth. Different elements in sea water is not directly proportional to the different elements in river water, but is inversely proportional to the elements in seawater that are insoluble. The concentration of an element whether it is in the ocean or in the lake depends on both input and output on the chemicals and its reactivity. Elements removed from seawater are incorporated into particulate matter. The particulate matter then settles on the ocean floor. Solid precipitation or absorption of inorganic or organic particles can take place. Most elements in seawater have lower concentrations than insoluble compounds. Precipitation, therefore, does not take place. Adsorption reactions control the concentrations of most of the species in seawater. Reactions at sediment surfaces also remove ions. Alkali metals, alkaline earth metals, halides, bicarbonates, and sulfates are removed from the oceans at a slower rate than transition metals and nutrient elements.

Gases

Both oxygen and carbon dioxide are involved in lakes and oceans. The distribution of oxygen is influenced by biological processes and the stability of the water column. Well-mixed surface waters exposed to the atmosphere have oxygen concentrations close to equilibrium. Photosynthesis occurs at surface waters at high rates. In effect, oxygen concentrations increase above their equilibrium values, and in deep water where respiration takes place gases exchange slowly with the atmosphere. The concentration of oxygen is below their equilibrium values. Reduced concentrations of oxygen in deep water is used to trace the flow of deep water that goes through ocean basins. Water that sinks in the deep ocean is saturated with oxygen. As water flows deeper and oxygen is consumed by microbes, the concentration of oxygen decreases. Oxygen concentration can be exposed as the apparent oxygen utilization or A.O.U. The A.O.U. is the difference between the saturation value and the measured concentration. The value of A.O.U. increases as oxygen content decreases.

The chemistry of carbon dioxide is different. Carbon dioxide dissolution has to do with ionic equilibria involving the carbonate species. Toxic organic pollutants can also affect both lakes and oceans. An example is polychlorinated biphenyls. These are distributed throughout many environments. They are found in air, water, rain, snow, soil and sediments. They are volatile. They enter and leave aquatic environments by transporting air across water-air interface.

Nutrients

Plants and animals also influence the concentration of chemicals in aquatic environments. Nutrient elements such as carbon, hydrogen, oxygen, and nitrogen are interconverted between dissolved organic and inorganic forms. Chemicals become recycled in oceans and lakes and are lost from the water column to sediments. Uptake and the release of nutrients take place by the process of photosynthesis and respiration. It is possible that there could be strong correlations between dissolved concentration of nutrients in lakes and oceans and that in organic matter. Finding ratios of nutrient elements are called Redfield Ratios. Elements can also be classified according to their vertical concentration profiles. They can be classified as biolimiting, biointermediate, and biounlimited. These correspond to species that are completely depleted in surface water as compared to deep water, partially depleted, or negligibly depleted.

The fraction of elements leaving surface waters by settling particles is represented as g. This is equal to particulate flux of chemicals in the deep ocean divided by the total flux of chemical. The fraction of chemicals carried to deep sea by particles equals their rate of loss to sediments divided by particulate flux is represented as f. The fraction of chemicals removed from the ocean equals the fraction of chemicals entering the deep ocean (g) times the fraction of particulate matter that is lost to the sediments in the deep ocean (f).

A phosphorus cycle takes place in lakes for phytoplankton. Phosphorus can enter and leave the lake like other chemicals. Phosphorus can settle to the sediments. It can remain in the dissolved form. The element phosphorus cycles itself between the eplimnion and hypolimnion layers of the lake. Bacterial decomposition and re-uptake of algae carries this process. Algae eventually settles in the hypolimnion and some phosphorus is lost to sediment by precipitating with iron and manganese phosphate minerals. Some phosphate in these minerals is released from sediments only if the deep water becomes anoxic. Other phosphorus releases back into solution when organic matter becomes decomposed by microbes with a small amount diffusing into the eplimnion. This is where it is regrouped by algae. Another small portion of phosphorus stays in the hypolimnion until the lake mixes.

Mixing and Reactions

Chemicals enter lakes and oceans during chemical weathering. Dissolution of minerals takes place. Minerals get precipitated in solution. Minerals also get deposited in sediments. This is how chemicals get removed from the water and the atmosphere. Metal oxides, carbonates, and hydroxides have low solubilities. These minerals limit the concentration of metal species in natural waters. Sediments in which there is low oxygen and high organic matter, sulfides are produced by reducing sulfate. Metal sulfides that have low solubility will be formed. Places where evaporation leads to supersaturated solutions, chlorides, sulfates, and carbonates precipitate. Water and acids come into contact with minerals. Chemical reactions occur that leads to the dissolution of ions. These reactions can occur in ground water that can enter into rivers, lakes, and oceans. We can think of igneous rocks reacting with acid volatiles to give sedimentary rocks and oceans that contain salt. Acids emitted from the inside of the Earth react with the minerals in igneous rocks which results in ions being carried to the oceans. They can enter the composition of water or they can be deposited in sediments.

Rocks get broken down into smaller pieces. Wind, expanding frozen water, and plant roots penetrating into rocks cause large rocks to be broken down to smaller pieces. Water picks up dissolved species from land and returns them to rivers, lakes, or oceans. Chemical weathering takes place when water and acids react with rock minerals to release ions. Silicates is one of the most important weathered minerals. Higher temperatures and higher precipitation lead to higher weathering rates. Organic acids such as carbonic acid help dissolve minerals. Sulfuric acid and nitric acid also increases chemical weathering. The acids convert basic species to soluble forms. Organic acids act as chelates towards metal cations. Solubility increases and precipitation gets reduced. Primary minerals are minerals that undergo chemical weathering. Secondary minerals are minerals formed from the products of weathering.

Igneous rocks are formed by crystallization from high temperatures. Sedimentary rocks are formed by sedimentation in water of secondary minerals. Metamorphic rocks are formed by both igneous and sedimentary rocks recrystallized at high pressures and temperatures where there is no melting. Dissolution reactions can also be categorized as congruent or incongruent. Congruent dissolution is a process where a mineral dissolves in stoichiometric amounts. Incongruent dissolution is a process where dissolved elements precipitate again into secondary minerals in nonstoichiometric amounts. Secondary minerals change to primary minerals during chemical weathering. Rivers also play a role transporting water filled with suspended and dissolved matter. More chemicals from rivers enter into oceans than from the atmosphere. Rivers form from water being runoff by land. Chemicals can be found in rivers as inorganic and organic matter whether they are dissolved or not. Water from rivers enter the ocean where substances can be deposited in sediments or reman dissolved.

Chemicals that enter the ocean can be deposited as sediments. The chemicals that go in and out of this cycling process can tell us a lot about the ocean. The ocean floor is made up of three different types of sediment materials. Primary and secondary silicate minerals are formed and which makes up one type of sediment material. A red clay sediment covers the ocean floor. Calcium carbonate also forms another sediment, while Opal forms another sediment. Both sediments are formed by depositing phytoplankton and zooplankton shells. Chemicals added to the ocean take place by adding river water, aerosols being deposited, reactions involving volcanoes and seawater, reactions with minerals, and biological processes. Chemicals can be removed from the ocean by biogenic deposition of materials, evaporating minerals, transfer of sea and air reactions involving volcanic gases and seawater, minerals being removed, and water buried from beneath pores. Rivers are a major source of chemicals for both lakes and oceans. Oceans get their major source of water by rain. Sulfuric acid and nitric acid from rain can become absorbed into both cloud and rain. Aerosol particles can contribute quantities of trace metals to the ocean. Sea-spray carries sea salt particles on land. Salt is removed from the oceans for a short time.

Elements that are present in water can occur as particles or be dissolved. Ions dissolved in water are hydrated by water and can form complexes with other species. Particulate material attached to the surface are reactive and can be absorbed by metals and ligands. This affects how many ions are present in water. Reactions that predominate depend on the surface and the functional groups, the pH, metal ions, ligands, and other types of ions. Redox reactions with biological processes also take place in lakes and oceans. Trace metals are also present in natural water. Lakes have more trace metals than oceans. They are influenced by adsorption, complexation, and redox chemistry.

Conclusion

The atmosphere of the Earth makes up mainly oxygen and nitrogen. There are other trace gases present in the atmosphere. Solar radiation, erupting volcanoes, and the actions of society influence the chemistry that takes place in the atmosphere. Ozone absorbs ultraviolet radiation in the stratosphere. It also protects us on Earth. The use of chlorofluorocarbons have been eliminating the ozone layer. Eruptions from volcanoes reduce ozone in the stratosphere and climate becomes affected. Carbon dioxide, chlorofluorocarbons, methane, and other gases affect global warming. Sulfur dioxide and nitrogen oxides cause acid rain. Photochemical smog comes from the exhaust of automobiles when there is sunlight. Air pollution can also come from radon, carbon dioxide, and carbon monoxide.

The chemistry of our environment also takes place in rivers, lakes, and oceans. The chemical composition of natural waters is different from all three. Physical and chemical weathering takes place where dissolved ions can be transported to rivers, lakes, and oceans. Gases and atmospheric pollutants take place in all three of these systems. The thermocline is a region of decreasing temperature. Regions of increasing density is called the pycnocline. The surface and deep regions of lakes are called eplimnion and hypolimnion. They are separated by a region called the thermocline. Dissolved ions and precipitates occur in natural waters. Photosynthesis and respiration takes place where both oxygen and carbon dioxide are continuously exchanged. Oxygen concentration can be expressed as the apparent oxygen utilization or A.O.U. The A.O.U. is the difference between the saturation value and the measured concentration.

Nutrient elements such as carbon, hydrogen, oxygen, and nitrogen can be interconverted between dissolved organic and inorganic forms. Finding ratios of nutrient elements are called Redfield ratios. Elements can be biolimiting, biointermediate, and biounlimited. The fraction of elements leaving surface water by settling particles is referred to as g. This is equal to particulate flux of chemicals in the deep ocean divided by the total flux. The fraction of chemicals carried to deep sea by particles that equal the rate of loss to sediments divided by particulate flux is referred to as f. Phosphorus cycles takes place in lakes for phytoplankton.

Soluble minerals and precipitates form in natural waters due to physical and chemical weathering. Primary minerals are minerals that undergo chemical weathering. Secondary minerals are minerals from the products of weathering. Igneous rocks, sedimentary rocks, and metamorphic rocks form on planet Earth. Dissolution reactions can be congruent or incongruent. A mineral that dissolves in stoichiometric amounts is referred to as congruent dissolution. Dissolved elements that precipitate into secondary materials in nonstoichiometric amounts are referred to as incongruent dissolution. The ocean floor is made up of different sediments. These sediments include primary and secondary silicate minerals, calcium carbonate, and opal. Trace metals can also be present in natural waters. Input and output of chemicals take place by various processes such as evaporation, deposition, and other processes. Reactions that predominate depend on the surface, the functional groups, the pH, metal ions, ligands, and other types of ions.

Chapter 22

s and p block Elements

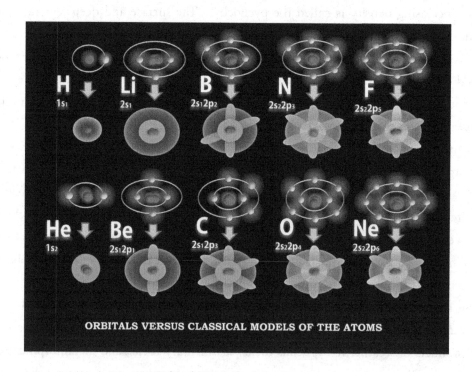

Orbital Models of the Atoms

Hydrogen, an odorless and colorless gas, is the most abundant and lightest element in the universe. It is in space, and it is one of the elements in stars. There is very little hydrogen in the Earth's atmosphere. It is found in water molecules that make up our oceans and seas. Applications of it include rocket fuel, the hydrogenation of fats, and petroleum desulfurization. Helium, the second lightest element, is an odorless and colorless inert gas. It exists primarily in its pure element form. Helium comes second as the most abundant element in the universe. It is scarce on Earth, and it makes up one of the gas components in air. It is used for balloons, blimps, and as a nuclear plant coolant. It is also used for leak detectors. Lithium is a soft and silver-white reactive metal. It is never in its pure state since it reacts quickly with air and water. It is found in the mineral Spodumene. This is where lithium can be extracted from compounds and stored in oil or kerosene. It is used in batteries for pacemakers. It can be combined with other elements to make an alloy used in space. Beryllium is steel-gray, strong, lightweight, and brittle. Several forms of it exist in Earth's crust. It is too reactive to be found in its free state. Its natural source is from the mineral Beryl. It is used for X-ray tube windows. It can also be combined with other elements to make watch springs. It is also used in compounds for heat conducting ceramics. Boron can either be in the form of a brown powder or hard and brittle crystalline black. It is a scarce element found on Earth, and it is never found in its pure form. It is used as a dopant for semiconductors, and it has become important for the manufacturing of transistors. It is also combined with other elements to make tennis rackets and heat resistant glass. Carbon can be diamond clear or black as found in graphite. Almost all of plants and animals are made of carbon. It represents a very small percentage of the Earth's crust, but it is the most abundant element found in nature. It is found in coal, petroleum, perfumes, carbohydrates, proteins, and enzymes. Carbon is also used to make plastics, pencils, and steel. Nitrogen is a colorless and odorless inert gas. Nitrogen is the main component of gas in the Earth's atmosphere, and it is in greater amount than oxygen. Compounds of nitrogen are important for agriculture and industry. It is used for cryogenic surgery, to make ammonia, and it can also be a liquid used as a coolant. Rocket fuels contain compounds of nitrogen too. Oxygen is a colorless gas, and it can be seen as a pale blue liquid. Less than half percent of Earth's crust is made up of oxygen. It is used for combustion of fuels that provides us energy and heat, making steel, and to purify water. Oxygen is also found in our atmosphere that is important for our survival. It is replenished by photosynthesis in plants. Fluorine is a very pale yellow gas, but as a liquid, it is bright yellow. Fluorine is too reactive to be in its elemental form. A lot of it is found in minerals such as Fluorite and Fluorspar. Small quantities of it is found in teeth, bones, blood, and seawater. It gets converted to fluoride that is used in steel making, aluminum refining, toothpastes, and water fluoridation.Neon is a colorless, unreactive, and inert gas. It does not form any compounds with other elements. It is also found in our Earth's atmosphere. An extremely small percentage of it is found in air. It emits an orange-red glow in a discharge tube. It is used in television tubes, lasers, and detectors using voltage. It is also seen as fog and neon lights.Sodium can be described as a silvery-white metal. It is a highly reactive metal that does not occur by itself in nature. It is normally immersed in kerosene to make sure it does not react with air or moisture. Sodium is also present in the Earth's crust. It is also found in an alloy as a coolant for nuclear reactors and batteries. Sodium is used for streetlights. Magnesium is a shiny gray solid. It is lightweight, reactive, and malleable. The free element is not found in nature. It comes from the mineral Magnesite. It is also present in the Earth's crust and in large quantities in seawater. Magnesium has been important as a structural material. Bikes are made up of an alloy containing the element magnesium.Aluminum is silvery, gray, and has a metallic appearance. It is the most abundant metal in Earth's crust. Aluminum comes from the ore Bauxite that is found all over the world but mainly in tropical regions. Aluminum is known for its commercial importance. It is used for window frames, foil, and doorknobs. It can be found in alloys used for cars and airplanes.Silicon looks crystalline with a reflective appearance. Some of the crystalline faces have an interesting light blue appearance. It is the second abundant element found in Earth's crust. Silicon is used for microchips and solar cells. Silica is a compound made up of silicon. It has a crystalline form called Quartz and a noncrystalline form called Flint.Phosphorus can be colorless, white, yellow, scarlet, red, violet, or black. The human skeleton and teeth contain calcium phosphate. In nature, it is found in the form of phosphate rocks and Fluoroapatite. Phosphorous is used commercially to make lots of phosphoric acid for fertilizers. It is also found in compounds used for toothpastes and detergents. Sulfur is a yellow solid that looks like chunks of crystals. It is reactive in its free state and in ores and minerals. A small percentage of it is found in Earth's crust by mass. Sulfur produced commercially comes from underground deposits. It is found in matches, used for fireworks, batteries, mixed with polymers to make rubber, and in compounds for hair lotion.Chlorine is a pale yellow-green gas. It is found in dissolved salts found in seawater and in deposits such as salt mines. Chlorine is also used to purify water and mixed with other compounds to bleach and to remove stains. It is also found in compounds used for plastics. Unfortunately, some chlorine compounds have caused environmental problems. Argon is a colorless and inert gas, but it has a violet glow when placed in a discharge tube. It is found as a small percentage in Earth's atmosphere. It can be obtained by distilling air. Although argon is a noble gas, compounds of it have been successfully formed. It is used for light bulbs and lasers. It is also found in Geiger Counters used in laboratories. Potassium has a silvery gray appearance. It is very reactive, and it is never found in its free state. It is also found in silicate minerals and in seawater. Potassium is also important for the growth of plants. It is used as a compound for fertilizer, matches, gunpowder, and glass lenses. It can form ionic salts, and it is found in nature and in laboratories. Calcium looks dull, gray, and silver that is a soft metal. It is also one of the most abundant elements in the Earth's crust. Our teeth and bones contain the element calcium. Calcium is also found in sedimentary, igneous, and metamorphic rocks. It is also found in compounds such as calcium carbonate, which is used for chalk, marble, and limestone. Gallium looks silver-white. Gallium is used in quartz thermometers. It has an unusual ability to expand when it freezes. It is commercially made as a by-product when refining other metals such as aluminum. Gallium is also used in compounds to locate tumors. A compound of gallium arsenide is used as a laser dioide that converts electricity into a ray of laser light. Germanium, a rare element, looks grayish-white. It is found combined with oxygen to form a mineral. Germanium is a semiconductor, and trace amounts of it can increase the conduction of electricity. Doped germanium is then used to make transistors. Germanium is also used in wide-angle lenses, infrared night vision and in compounds for fiber optics. Arsenic, a metallic gray element, is a poor conductor of electricity. Most of it is found in minerals such as Realgar and Orpiment. Arsenic has become important in the field of solid-state electronics. A small amount of them is added to semiconductors to make them into transistors. It can also be made with other elements to make light emitting diodes. Selenium has two different forms, and it can either be black or red. A lot of it is recovered commercially as a by-product from refining copper and making sulfuric acid. It has become a useful conductor in the presence of light. It is also used in light meters, copy machines, and solar cells. Compounds of selenium have been found in dandruff and shampoos. Bromine, a red-brown color, exists as a gas or a liquid. As a solid, it has a metallic luster. Bromine is found in seawater, salt mines, and brine wells. It is found in compounds used for disinfectants and photographic film. Bromine is used to produce a gasoline additive called ethylene dibromide that removes lead additive after gasoline is combusted. Krypton is a colorless gas, and it shows a white glow when placed in a discharge tube. It can be separated from air by fractional distillation. It has an extremely small percentage of abundance in air. Its main use is in neon lights. It is also used for fluorescent bulbs and flash bulbs when mixed with the element xenon. It is also used for ultraviolet lasers. Rubidium looks gray-white. The metal itself is stored in kerosene. Rubidium is obtained as a byproduct when refining lithium and cesium ores. This element has not been used very much. The element is made in making television and cathode-ray tubes. As an element, it can also be used to scavenge gases in vacuum tubes. It is also used in heart muscle research. Strontium looks metallic and silvery-white. It is found from the mineral Celestite and Strontianite. Salts of strontium give a red color when they burn in air. They are used in highway flares and also seen in fireworks. It is found in compounds that are used in nuclear batteries in buoys and in phosphorescent paint. It is also a source of radiation for beta particles.Indium is a silvery and lustrous gray rare element. It is very corrosion resistant. Most of the supply of indium comes as a by-product when zinc is refined. It is mainly used in alloys of other metals. Alloys containing indium conduct heat better. It is used in solar cells and mirrors. It is also used in electronics as photocells and transistors. Tin is a silvery or gray rare element. It is present in extremely small amounts in Earth's crust. It is found in the ore Cassiterite. Tin is very malleable. It is mainly used as an alloy to make tin plate. It is found in mixtures, which is used in coins, cups, and plates. It is also used in organ pipes as an alloy. Bronze, solder, and pewter are alloys that contain a certain percentage of tin. Antimony is silvery and lustrous gray. It is a poor conductor of electricity. It comes from the mineral Stibnite. Antimony is used for safety matches. The head of the match has a mixture of antimony compounds and other compounds too. It is used to make compounds used for ceramic glazes. It can also be used as a compound to make fire retardants. Tellurium, a silvery and lustrous gray rare element, can combine with gold to form a telluride. Very few elements can combine with gold. It is one of the components of ores that contain gold. Tellurium can be recovered as a by-product when gold is refined. It is used for percussion caps and for a battery plate protector. It is also vulcanized with rubber.Iodine looks like a violet gas that is lustrous and nonmetallic. It is found in seaweed and wells that contain brine as a salt or as an organic iodide. Iodine is recovered from sodium iodate that is found in saltpeter or potassium nitrate. Iodine is a disinfectant, used for halogen lamps, and it is found in compounds, which is used for ink pigments. Xenon is a colorless gas. It is found in the atmosphere in trace amounts. It has no color, no odor, and no taste. This element is a lot heavier than air. Xenon can be extracted from air by fractional distillation. It shows a blue glow when placed in a discharge tube. It is used for ultraviolet lamps and sun lamps. It is also used for projection lamps.Cesium is silvery gold. It is the softest metal known. It comes from the mineral Pollucite. It can act as a scavenger by reacting with unwanted gases in a television tube. As an element, it is used for photoelectric cells. Scintillation detectors contain cesium iodide crystals. The crystals get radiated producing pulses of light which converts to electrical signals. Barium is silvery and gray. It is abundant in the Earth's crust. It is found in the minerals Witherite and Baryte. As one of the elements in an alloy, it can be used to make spark plugs. Zinc sulfide and barium sulfate mixed together produces a white paint pigment called Lithopone. Compounds containing barium is used to make green fireworks. Thallium is a silvery-white carcinogenic metal. It is scarce and malleable. It is found in ores such as Crookside, Lorandite, and Hutchinsoninte. It is also found in manganese nodules on the ocean floor. The element is normally obtained, as a by-product when lead and zinc is refined. It is also found in compounds used for heart muscle research.Lead is a metallic gray metal. It is very malleable, and it is a poor conductor of electricity. It is very hard to find the element by itself in nature. Lead is obtained from the ore Galena. As an element, it can be used to protect us from radiation. It is used to make electrodes, and it is a component of solder. Compounds of lead are also used as paint pigments.Bismuth is a lustrous, silver, heavy, and brittle metal that resists corrosion. It can be found in its metal state in nature. It can also be found in its ores Bismite or Bismuth Glance. Bismuth is obtained as a by-product when copper, tin, and lead is refined. As an element, it is a catalyst for rubber production. It is also found in antiacids and antidiarrheals. Polonium is a silvery rare metal. It comes in extremely small trace amounts from the uranium mineral Pitchblende. This element can be formed through radioactive decay of uranium and thorium. Applications of this element involve mainly its radioactivity. As an element, it is found in nuclear batteries. It is also used as a photographic film cleaner. No one knows the appearance of Astatine. But it is probably a black solid and possible metallic. It is a member of the Halogen Family. It is used for alpha particle radiotherapy to treat some diseases. It is a radioactive element that does not last long. Small amounts of this element can be produced from uranium and thorium radioactive decay. Radon is a colorless gas. It is formed when uranium and thorium decay into its by-products. It is the heaviest gas, and it is much more dense than air. It belongs to the Nobel Gas Family, and it is not very reactive. As an element, it is used to predict earthquakes. It has been made for radiation therapy and to treat cancer. Hopefully, this element can treat other problems. The appearance of Francium is not known. But it is possibly metallic. It is considered the heaviest of the Alkali Family. This element is made artificially in nuclear reactors. The element itself is not very stable. It is produced by radioactive decay from uranium and thorium. A sample of Uraninite contains an extremely small specific mass of francium.Radium is silvery-white metallic. It comes in extremely small trace amounts from the ore Pitchblende. It is the last member of the Alkaline

Earth elements. It is found in compounds used for glow in the dark paint. In medicine, radium is in the form of a salt called radium chloride. It is also used at medical facilities to make the gas radon to treat cancer therapy.

<u>Conclusion</u>

The Periodic Table of Elements is a chart that maps out the chemical elements arranged based on its atomic number. The atomic number tells us its specific identity. Some of the elements have been discovered since ancient times and through the years up to the 1700s. The elements existed, but they were not grouped or classified systematically because there was still a lot more to learn about them. In the late 1700s, some of the elements discovered were grouped with basic properties whether it was a metal, nonmetal, or in between. During the 1800s, many chemists working with new elements being discovered at that time grouped them into triads. There were many different triads, but there was still much work to carry out and group them all together. A remarkable discovery came about when the elements could be arranged based on other properties. Much more elements could then be arranged closer together. Elements were then arranged based on increasing atomic weight measured in atomic mass units or abbreviated as amu. For each eighth element written, it was thought the elements followed a recurrence of eight known as The Law of Octaves. However, this idea proved to be incorrect. During the mid to late 1800s, the Periodic Table of Elements was finally arranged into bigger vertical columns and bigger horizontal rows. Although incorrect during that time period, they were arranged based on increasing atomic weight. During this time period, there were still gaps of missing elements on the Periodic Table. Chemical trends were noticed and some of the elements were soon arranged closer together neglecting whether the atomic weight was bigger or smaller. The elements were then correctly arranged based on atomic number. In the early 1900s, many chemists discovered new elements that filled in the gaps on the Periodic Table. Extremely small quantities of more new elements were also made in laboratories. These synthesized elements are not found in nature and were added to the Periodic Table of Elements. There are still more elements being added to the Periodic Table of Elements. The Periodic Table of Elements is now organized based on the number of rows and columns. There are seven consecutive rows of elements carefully arranged by scientists. The elements are also arranged in groups or families based on its vertical column. There are eight main columns total on the left and ride side of the Periodic Table that correspond to properties for each of the elements. There are twenty-four small vertical columns in the middle of the periodic table. These correspond to elements that have similar chemical properties too, but there are still other properties for each element completely different. The elements in Group IA on the Period Table are called the Alkali Metals or the Lithium Family. Group IIA is called the Alkaline Earth Metals or the Beryllium Family. Both of these groups are almost similar on the periodic table. These elements are all metals with the exception of hydrogen located at the very left hand corner, sometimes in the middle, and sometimes placed as the second to the last element before helium on the periodic table. Moving towards the right hand side of the larger portion of the periodic table we find that Group

IIIA is called The Triels or the Boron Family. The next vertical column is Group IVA. This is called The Tetrels or the Carbon Family. Group VA is called the Pnictogens or the Nitrogen Family. Group VIA is called Chalcogens or the Oxygen Family. Group VIIA is considered to be the Halogens or the Fluorine Family. Finally, Group VIIIA are the Nobel Gases, the Helium Family, or the Neon Family. The elements in Groups IIIA to VIIIA together form this block on the periodic table.

Chapter 23
d and f block Elements

Evolving Chemistry

Scandium is a silvery-white metal. A very low percentage of it is present in Earth's crust. A lot of it is found in the sun and other stars. It is a lightweight material, and it does a good job resisting corrosion for aircrafts. It is also used as an alloy for large screen televisions and stadium lighting. Alone as an element, it can also be used for leak detectors. Titanium is a silver-white and gray ductile metal. It is also present in the Earth's crust. It comes from the minerals Rutile and Ilemnite. Titanium is considered lightweight, but it is very strong and resistant to corrosion for rockets and jet engines. Titanium can be used as a heat exchanger and as an alloy such as bone pins and pigments for paint and paper. Vanadium is a blue, silver, and gray metal. A small percentage of it is found in Earth's crust. It is very resistant to corrosion. The element itself comes from the mineral Vanadinite. Vanadium is an additive to making steel stronger than it is. It is normally used as an alloy for tools and construction materials. It is also used as an alloy for jet engines. Chromium is silvery metallic. It is brittle, but it is also a hard material. The metal is extracted from Chromite. It is used for plating car parts such as bumpers. It is also used to make stainless steel. It is found in compounds used in stereos and videotapes. One of its main uses as a compound is the making of colored pigments and trace amounts of its color for gems. Manganese looks silvery metallic. It corrodes in moist air. It does not exist in its free state, but it is found in the mineral Pyrolusite. Another source of manganese are the manganese nodules at the bottom of a sea. It is an alloy that is used for plows and batteries, and it is used as an alloy for stainless steels. It adds strength and resists corrosion to certain alloys. Iron has the appearance of lustrous metallic. It is also present in Earth's crust. The core of the Earth is known to contain mainly molten iron. It is not found in its pure state in nature. Hematite and Magnetite are minerals containing iron. Iron is normally made in a blast furnace. It is used to make bikes, cars, and bridges, and it is used as an alloy to make magnets and machines. Cobalt is hard, lustrous, and a gray metal. A small percentage of it is found in the Earth's crust. It comes from the ore Cobaltite. Cobalt is obtained by roasting Cobaltite in air. It is also added to steel to reduce its resistance from corrosion. It can be mixed with other metals to form alloys. It can be used as a gamma radiation source, and it is used as an alloy for permanent magnets. Nickel looks lustrous, metallic, and silver. It comes from the ore called Millerite. It is a rare element consisting of a small percentage of it on Earth's crust. The molten core of the Earth contains nickel. Coins are made up of the element nickel. It can be made into an alloy for knives, forks, and spoons. An application of it is the nickel-cadmium battery. Copper has a red-orange metallic luster. It is a rare element making up a very small percentage of it on Earth' crust. It comes from the ore Chalcopyrite. It can be obtained by roasting the ore or by electrolysis. It is one of the best conductors of electricity. Copper is used for cables, wires, water pipes, and it is also blended in with other elements to make pennies. Zinc looks silver-gray. It is not an abundant element and makes up a very small percentage on Earth's crust. It comes from the mineral Sphalerite or Zincblende. Zinc tends to form an oxide with air making it less reactive. It is used in batteries, gutters, and in compounds to make white pigments for rubber. It is also found as an alloy in water and gas valves. Yttrium looks silvery-white. Only small quantities are found in the Earth's crust. Rocks taken from the moon has a high amount of Yttrium content. Most of it is obtained from Monazite Sand. It is found in compounds used for color television screens and radar. It is also blended to make compounds used for superconductors. It can also be used for solid-state lasers. Zirconium looks silvery-white. It is a strong metal that is resistant to both corrosion and high temperature. It is used mainly in the making of space vehicle part supplies. A compound of zirconium comes from the mineral Zircon. As an element, it is used for nuclear fuel rods and catalytic converters. It is also blended in with other elements to make gemstone. Niobium looks gray, metallic and bluish when it is oxidized. It is found in the mineral Columbite. The element has been placed as high importance for high temperature superconductivity. Niobium is also added to steel for materials to withstand high temperatures. It is used as an alloy for welding rods and cutting tools. It is also used for pipelines. Molybdenum is gray and metallic. It is mined from the ore Molybdenite. Molybdenum is added to steel that can be used for automobiles and engines for aircraft parts. This is important because it can withstand temperature and pressure changes inside an engine. As an element, it is used as a filament in electric heaters. It is also a source of radioisotopes in hospitals. Technetium looks like a shiny and gray metal. It is not an element found in nature on Earth. Trace amounts have been discovered in some stars. An unstable compound of technetium can be injected in the veins of a patient. A photographic plate will reveal how those organs are working. It is also used as a radiation source in medical research laboratories. Ruthenium is a silvery-white metal. It is recovered as a by-product when platinum ores is refined. It is used as a catalyst for industrial processes. As an element, it is used for eye treatment. It is also used for meters to measure the thickness of eggshells. It is also used as an additive for platinum in the jewelry industry and added to titanium to help resist corrosion. Rhodium is silvery-white metallic. It is a rare element and found with the ores platinum. It is used as an additive to make platinum hard and as a catalyst in the industry. In the automobile industry, it is used as headlight reflectors and catalytic converters are filled with catalytic beads of rhodium. Telephone relays have also been used that contains rhodium. Palladium is silvery-white. It is easily malleable and ductile. It is also used as a metal catalyst for hydrogenation of organic compounds, and it is used in catalytic converters. It is used in mixtures to make dental crowns and jewelry because of its fine resistance to corrosion. Some palladium compounds are used as anti-tumor agents. Silver, a lustrous white metal, can be found in nature as its free state. It is found in ores such as Argentite and copper-nickel ores contain it as an impurity. It is used in mirrors and batteries. It is also combined with other elements to make an alloy used for silverware. Silver is used to make photographic film and paper. Above all, silver is the best conductor for heat and electricity. Cadmium is silvery and bluish-gray metallic. It is a rare element found in Earth's crust. It is found in ores such as Cadmium Sulfide. A lot of cadmium can be collected from zinc ores as a by-product when zinc is refined. It is also used for rechargeable batteries, plating of screws and bolts, and it is used mainly for electroplating of steel to make sure it does not corrode. Lanthanum is silvery-white. It makes up the first element in the Rare Earth Series. It is ductile and malleable. It comes from the ores Monazite Sands and Bastnasite. It is used as an alloy to make lighter flints, and compounds for camera lenses. Lanthanide compounds is used in making electrodes for carbon arc lamps, for studio lighting, and movie projectors. Cerium is silvery-white, and it is the most abundant of the metals of the Rare Earth Series. It is ductile and malleable. It comes from the ores Monazite and Bastnasite. As an oxide, it is used for self-cleaning ovens. It is used to make electrodes for carbon arc lamps for searchlights and projectors. Cerium oxide compounds is also used to polish lenses for cameras. Praseodymium looks metallic. It is found in Monzanite and Bastnasite ores. It is malleable and ductile. Praseodymium oxides are used to make the electrodes of carbon arc lamps for searchlights and motion-picture projectors. Alloys of praseodymium are used for automobile and aircraft parts. Salts of praseodymium are added to enamels to make them look yellow. Neodymium looks silvery-white. The main ores of neodymium is Bastnasite and Monazite. The element and its oxides are good for making colored glass that is used for an artificial ruby for lasers. It is also used as a strong magnetic alloy for disc drives. The element in a disk magnet is powerful enough to detect the ink between real and fake money. Promethium looks metallic. This element is made in nuclear reactors. It is not found on Earth's crust. It is found in compounds used as a starter for fluorescent lights. It is also found in compounds used for nuclear batteries. Salts of promethium have been made. The radiation this element emits gives off a blue light in the dark. Samarium looks silvery-white. It is a rare earth element found in Earth's crust. The main ores for Samarian is Monazite and Bastnasite. It is found in compounds used for ceramic capacitors and for high temperature permanent magnets. Oxides of this element, found in certain glasses, have also been used as an absorber of infrared radiation. Europium is silvery-white. This is rare of all rare earth elements. The main ores for this element is Monazite and Bastnasite. Europium is ductile. It is found in compounds used to improve the red phosphor in color television tubes and computer monitors. It is also found in compounds for trichromatic fluorescent lights that save a lot of energy. Gadolinium is silvery-white. It is a rare element found in earth's crust. The main ores for this element is Monazite and Bastnasite. It is ductile, malleable, and strongly attracted to magnets. It is found in compounds used for diagnosing osteoporosis. It is also found in compounds used to make phosphor for color television tubes and computer memory. Terbium is silvery-white. This rare earth element is not very abundant. It comes from the ore Monazite. It is malleable, ductile, and resists corrosion. Terbium compounds have been used as phosphors that give the green color in television tubes and computer monitors. It is combined with other elements to make a magneto-optic alloy used for compact discs. Dysprosium is silvery-white. It is a rare earth element found in the Earth's crust. The main ores come from Monazite and Bastnasite. It is combined with other elements to make a magneto-optic alloy for compact discs. It is found in compounds for color television tubes. Dysprosium is used in control rods for nuclear reactors. This is under consideration. Holmium is a silvery-white rare earth metal. Small quantities of it exist in Earth's crust. Obtaining this element in its pure form is now available due to sophisticated modern instrumentation. It comes from the ore Monazite. It is malleable, ductile, and resists corrosion. It is found in compounds for glass coloring. It is also found in compounds for eye-safe lasers. Erbium is a silvery-white rare earth metal. The main sources of erbium are from the minerals Xenotime and Euxerite. It is considered as an impurity in these ores. It is malleable and resists corrosion. Oxides of erbium are added to glass to make them look pink. It is also found in compounds for coating of sunglasses and jewelry that is not expensive. Thulium is silvery and gray. It is a rare earth metal that is very scarce. This element comes from the mineral Monazite. It is ductile, malleable, and resists corrosion in air that is dry. It is found in compounds used for lasers. Remarkably, the element has also been used as an X-ray source and as a high temperature superconductor. Ytterbium is silvery-white. It is a rare earth metal found in an ordinary abundance in Earth's crust. It is malleable and ductile. It comes from the mineral Monazite. It can be used as an alloy to improve the strength of steel. As an element, Ytterbium is a portable

X-ray source for blood treatment. It is also found in compounds as dentures. Lutetium is silvery-white and makes up the last element in the Lanthanide series. It is not very abundant on Earth's crust. It comes from the mineral Monazite. It can also resist corrosion. It is not easy to prepare this metal in its pure form. I t is found in compounds used for temperature sensing optics, and it can also be used as a catalyst for petroleum cracking in refineries.Hafnium is steel gray. It is very ductile and very resistant to corrosion. This element is obtained from the ores Zircon and Baddeleyite. Nuclear submarines have reactors that contain hafnium control rods. It is also used to control nuclear reactions. It is used in incandescent lamps, and it removes unwanted gases from a system. Tantalum is a gray-blue heavy and hard metal. It comes from the mineral Columbite. As an element, it is used for capacitors. It is also used as an alloy for weights. Tantalum can be used to replace hip joints. It can be used as a plate to replace damaged parts of the human skull. Screws and staples of this element can hold fragments of broken bones. Tungsten is grayish, white, and lustrous. It is known to be brittle. The main ores of tungsten are Wolframite and Scheelite. It is also used for lamp filaments. Filaments of tungsten are found in television tubes and in cathode ray tubes inside computer monitors. It is also used in electric heaters, nozzles for space vehicles, and compounds for cutting tools. Rhenium is a silvery-white rare element found in nature. It comes from the ores Molybdenite and Copper Sulfide. It is mainly used as an alloy for making metals that can be used without it being worn out. Electrical switch contacts and electrodes contain alloys of rhenium. It is also found in alloys used for thermocouples and in compounds for oven filaments. Osmium is silvery, blue cast. It is found mainly in nickel and platinum ores. It is used to make hard alloys with platinum and iridium. It is used as alloys for electrical switch contacts, fountain pen points, and compass needles. A compound called osmium tetroxide in dilute solutions can be used to stain substances to be seen on microscope slides. Iridium is a silvery-white brittle metal that resists corrosion. This element is much more denser than liquid water. This element is also found in platinum and nickel ores. Iridium is added to platinum to make hard alloys. As an element, it is used for satellite thruster engines and hypodermic needles. It is also found in alloys used for helicopter sparkplugs. Platinum is a grayish-white metal. It is used to make jewelry. It is malleable, resists corrosion, and found in nature as the metal itself. It can be used as a catalyst in certain reactions. As an element, it is used for crucibles. It is also found in mixtures for dental crowns and petroleum refining. Some platinum compounds can inhibit the growth of cancerous tumors. Gold is metallic yellow. This element has to be one of the most precious metals known. It is resistant to corrosion. It can be found in nature as nuggets, flakes, or minerals called Tellurides. Gold can normally be found near deposits of Quartz and Pyrite. It is found in alloys used for jewelry. It is also found in mixtures used for dental crowns. Mercury is seen as a silvery element found in nature. It is the only metal that is liquid at room temperature. It does a good job conducting electricity, but it does not conduct heat very well. It comes from the ore Cinnabar. It is also called Vermillion. As an element, it is found in barometers and thermometers. It is also found in mixtures used for dental fillings. Actinium is a silvery-white metal that makes up the actinide series. It is an element formed by radioactive decay from both uranium and thorium. Only a few amounts have been made artificially. It glows in the dark. The element actinium occurs very little in nature from uranium ores. A sample of Vicanite might have one or two atoms of actinium. Thorium is silvery that can turn into black tarnish metal. It is found in some electronic devices, and compounds containing this element have been used for portable gas lamps. Monazite Sand has a small percentage of this element. It is a coating on filament wire. It is also found in compounds used for crucibles. Thorium could possibly be a good source of energy sometime in the future. Protactinium is bright, silvery, and a metallic luster. It is found in extremely small concentrations of the uranium ore Pitchblende. It has been impossible to get a photograph of this element. It is possible that the mineral Torbernite may have some atoms of it during different times. It is also possible to use protactinium as a chemical tracer in geology. Uranium is silvery, gray, metallic, and the heaviest element found in nature. Its name derives from the planet Uranus. It is found in minerals such as Pitchblende, Uranite, and Monazite Sands. As an element, it is used for breeder reactor fuel, and it acts a shield from radiation. It is also found in compounds used for glass coloring and for glazes for ceramics. Neptunium is silvery metallic. It was the first element to be made artificially. Trace quantities of this element are found in nature in uranium ores. Its name derives from the planet Neptune. It is beyond the planet Uranus. Neptunium has been made in nuclear reactors as one of the by-products from plutonium. Commercial applications and use of this element is still unknown. Plutonium is a silvery-white metal. It is made artificially. Its name derives from the planet Pluto. Special reactors produce plutonium. It can tarnish in air to make an oxide giving a yellow color. This element has made it useful for power sources that are far away or difficult to get at like a pacemaker for a heart. It is also used as a film cleaner. Americium is a silvery-white metal that is made artificially. It was named after discovering this element in America. This element is found in huge quantities in nuclear reactors. It is found in compounds used for crystal research. It is also found in compounds used for smoke detectors. Elements like this have saved our lives and valuables. Curium is a silvery metal that is used as a radionuclide. This element was named after chemists Pierre and Marie Curie. Compact units containing curium have been useful for pacemakers, navigational buoys, and space probes for space missions. Powering instruments to operate for long periods of time has been used for this element, and it is a source for specific particles. Berkelium is silvery. It is a target nuclide to prepare heavier elements. It was named after the city Berkeley in the state of California. The metal itself has never been isolated. It is very scarce. Since an extremely trace amount of the compound berkelium chloride has been produced, there might be some use of this element in the future. Californium is silvery. This element was made in a laboratory at the University of California, Berkeley. It was named after the state of California. This element can be taken into the field for analyzing layers of earth that contain oil or for mining for the elements gold and silver by experimentation. Specific demonstrations of this element can be shown in laboratories. Einsteinium is silver-colored. This element was named after physicist Albert Einstein. It is produced artificially in high power nuclear reactors and in scientific laboratories. It is favored for producing ultra heavy elements. It is also used as a calibration marker for the Surveyor 5 lunar probe. Perhaps, it is possible this element can be used again for lunar surfaces.

The appearance of Fermium is not known. The element was named after physicist Enrico Fermi. Most of it is produced from high-power nuclear reactors. An alloy of it is used to measure the enthalpy of vaporization of the metal itself. More research is needed to learn and understand both the physical and chemical properties of this element. The appearance of Mendelevium is not known. The element was named after chemist Dmitri Mendeleyev. Combining Einsteinium with specific particles makes Mendelevium. It is an element made artificially. At this time, there are no applications or use for this element. Radioactivity and solution chemistry have been studied for this element. The appearance of Nobelium is not known. The name of this element was named after chemist Alfred Nobel. It is probably more likely silvery-white or gray and metallic. It could cause a radiation hazard if a certain number has been produced. It can exist as a complex compound or complex species in solution. More information is needed to learn about this element. The appearance of Lawrencium is not known. This element was named after Ernest O. Lawrence. He invented a machine in the laboratory called the cyclotron. The only compound known up to this point is Lawrencium (III) Chloride. The element is found in fusion and hot fusion of nucleosynthesis and decay products. More information is needed to study this element. The appearance of Rutherfordium is not known. This is the first of the transactinide elements. The element is named after physicist Ernest Rutherford. The element has been used for hot fusion, cold fusion, and decay studies. The $RfCl$ molecule has a tetrahedral shape. The appearance of Dubnium is not known. This element is named after the town Dubna in Russia. It is an artificial element used to study cold fusion, hot fusion, and decay of heavier nuclides. An example would be dubnium pentachloride or $DbCl$. Other Dubnium compounds have also been made. More information is needed to learn about this element. The appearance of seaborgium is not known. The element is named after chemist Glenn T. Seaborg. The element has been used to study cold fusion, hot fusion, and its decay to other elements. An example would be seaborgium oxide or SgO. Other seaborgium compounds have been made. More information is needed to learn about this element. The appearance of Bohrium is not known. The element is named after physicist Niels Bohr. It is an artificial element. The element has been used to study cold fusion, hot fusion, and its decay to other products. An example would be bohrium oxychloride or $BhOCl$. More compounds might be made and its applications and uses are still waiting to be discovered. The appearance of hassium is not known. The element was named after the German state of Hesse. It is an artificial element. The element has been used to study cold fusion and hot fusion. An example would be hassium tetroxide or HsO. More compounds of this element might be made. More information is needed to learn about this element. The appearance of Meitnerium is not known. The element is named after Lise Meitner. It is an artificial element. The element has been formed by nucleosynthesis reactions. The element has decayed to other elements. An example would be $MtCl$. Other compounds of meitnerium have been made. More information is needed to learn about this element. The appearance of darmstadtium is not known. The element was named after Darmstadt, Germany. It is an artificial element. The element has been used to study cold fusion, hot fusion, and its decay products. An example would be $DsCl$. Possibly more compounds could be made for this element. More information is needed to learn about this element. The appearance of Roentgenium is not known. This element is named after Wilhelm Roentgen. It is an artificial element. The element has been used to study cold fusion and its decay products. An example would be RgF or Roentgenium(III) Fluoride. The appearance of copernicium is not known. This element is named after the astronomer Nicolaus Copernicus. It is an artificial element. The element has been used to study cold fusion, hot fusion, and its decay products. An example would be CnF. Other compounds of copernicium have been made. More information is needed to study this element. The element ununtrium is a temporary name. This is an artificial element. It has been used to study cold fusion, hot fusion, and its decay products. A possible compound would be UutO. There are other possible compounds containing this element. The International Union and Applied Chemistry is still deciding on the name of the element. The appearance of Flerovium is not known. The element is named after physicist Georgy Glyorov. The element has been used to study cold fusion, hot fusion, and its decay products. The compound FlF is soluble in water. Other possible compounds have been made such as oxides and halides. More information is needed to study this element in more detail. The element Ununpentium is not known. This element has a temporary name. The element has been used to study hot fusion. It could form the compound UupCl. Other possible compounds of this element could involve other halides, sesquioxides, and chalcogenides. The International Union and Applied Chemistry is still deciding on the name of the element. The appearance of Livermorium is not known. This element is named after an American rancher Robert Livermore. It is an artificial element. This element has been used to study cold fusion, hot fusion, and its decay products. The compound LvO could be performed. Other compounds may involve combinations of other halides. The element Ununseptium is a temporary name. It is an artificial element. Studies of this element have been shown to predict properties based on its stability and other factors. The simplest compound is UusH. Other compounds have been predicted. The International Union and Applied Chemistry is still deciding on the name of the element. The element ununoctium is a temporary name. It is an artificial element. Predicted compounds have been proposed as well as its geometry. It is thought that this element is considered to be a solid instead of a gas even though it is in Group VIII: The Nobel Gas Family. The International Union and Applied Chemistry is still deciding on a name for this element.

Conclusion

The elements in the middle of the Periodic Table are called the Transition Elements. Group IIIB is the Scandium Family. This group consists of the Lanthanide and Actinide Series. Group IVB is the Titanium Family. Group VB is the Vanadium Family. Group VIB is the Chromium Family. Group VIIB is the Manganese Family. Group VIIIB consists of the Iron Family, Cobalt Family, and Nickel Family. Three families exist for this group. The numbering system then changes to IB. Group IB is called the Coinage Metals or Copper Family. Finally, Group IIB is called the Zinc Family. This middle block of elements including the Lanthanide and Actinide series form a wider block on the Periodic Table. The two rows of elements, known as the Lanthanides and Actinides, form another block on the Periodic Table. The Lanthanides consist of the elements Lanthanum to Lutetium. Likewise, the Actinides consist of the elements Actinium to Lawrencium. A brief description of each of the elements and its applications are mentioned. Each of the elements have a name and a symbol. Upgraded instrumentation and technology has allowed us to measure atomic mass units or amu to more decimal places. Let us look at each of the unique elements. As we study each of the elements, we will find that no one element is the same, and that each one differs from each other based on certain physical and chemical properties. It is possible for the information presented here can change in the future as science and technology continues to expand. More elements are still waiting to be discovered. There will be a new row of elements between the Alkaline Earth Metals and the Lanthanides and Actinides. Two rows of elements will make the Periodic Table of Elements wider than it already is. They might be called the Superactinides, and possibly the Superlanthanides. The chemical and physical characteristics of each of the newly formed elements might be intermediate between the Alkaline Earth Metals and the Actinides and Lanthanides. Once these properties are known and established, new compounds and molecules can be made. Many new applications of compounds and molecules are waiting to be discovered and made. Each of the chemical elements has been a building block for the creation of all the materials in the universe. Our life in the modern world has improved tremendously since ancient civilization. But we still have a long way to go to solve the challenging issues such as curing and preventing diseases, increasing our pure water supply, or even stopping global warming that face us today. The possibilities are endless. With advanced technology, we may be able to live in an enhanced world using each of the elements. New elements are waiting to be discovered for our use in society. Many of the man-made elements starting with neptunium have many applications still waiting to be discovered and learned. The problem is there are not enough of the elements with a sufficient mass and stability to understand more about the chemistry of them. We might be able to understand the advantages and disadvantages of each of them. The elements will continue to grow as long as there is more to learn in the changing world of chemistry and in many other scientific disciplines.

Chapter 24
Chemistry of Art

Wax Crayons and Measuring Flasks

Conserving art or restoring paintings is critical in order to preserve our cultural heritage and to maintain the beauty of an art piece. Chemical reactions take place on a microscopic level that lead to macroscopic changes observed as aging continues. Artists use a variety of pigments in their paints to give the color they desire. Art collectors and museums know temperature, intensity of light, relative humidity, and the quality of art is important. Origin, purity, combination of pigments, and the binding medium are a few factors that influence an art painting. Chemical and physical changes of pigments can take place to discolor the paintings even more if other factors are involved. Let us take a look at color since this is what first attracts many people.

Color can be considered an expressive element of design. It can affect our emotions directly and indirectly. Light can be considered a source of color which we can refer to as the electromagnetic spectrum. We can think of visible white light as a source of color. The color that we see for an object is based on the light source as well as the reflective and absorbing properties of the surface that the light strikes. For example, if we were to shine white light on a red cherry, all of the visible colors are absorbed except red. The red that is reflected is what we see. Color can begin where there is a source of light. If there is little light, there is little amount of color. When the light is strong, the color is more likely to be intense. During dusk or dawn, the amount of light is weak. It is difficult to distinguish one color from another. Areas where there is bright strong sunlight are known to be intense. Light originally comes from the sun. Atoms and molecules store and emit light energy during a chemical change. Light also occurs when excited atoms release energy. Warm colors include red-violet, red, red-orange, orange, orange-yellow, and yellow. Cool colors include yellow-green, green, green-blue, blue, blue-violet, and violet. Warm and cool colors help create a mood or an atmosphere. Warm colors show happiness, while cool colors show cold and peacefulness. In fact, our emotions we experience when we see colors is a result of our range of experiences. The range of experiences are called color conditioning. Color has also been thought of as having a symbolic significance. For example, white can stand for purity and light. Black can represent the absence of light. Yellow can be thought of as happiness or cheerfulness. Red can be thought of as energy, strength, and courage. Blue can suggest loyalty, wisdom, reliability, or justice. Violet can represent wealth, power, passion, or perhaps leniency. Green can be thought of as growth and life. Orange is shown as happiness, heat, or maybe glory.

Energy that travels in waves is called electromagnetic radiation. Frequency and wavelength of light are related to each other. Frequency times wavelength gives us velocity. We can think of velocity as the speed of electromagnetic radiation of waves inside a vacuum. The value is constant, and it is called the speed of light, symbolized as c, which is equal to about 3.0×10^8 m/s. Knowing either the wavelength or the frequency, one can find its other value using the equation $v = c/\lambda$ where v is the frequency, c is the speed of light, and λ is the wavelength. The sun gives off light which is in the form of waves that radiate energy. We can think of waves as disturbances that carry energy through space. Waves travel at the same speed in a vacuum, but they differ in wavelength and frequency. Some of these waves cannot be seen. Both visible and invisible waves are called electromagnetic radiation. The electromagnetic spectrum consist of gamma rays, X-rays, ultraviolet radiation, visible radiation, infrared radiation, microwaves, and radio waves. Gamma rays have the shortest wavelength and highest frequency and radio waves have the longest wavelength and lowest frequency. The visible spectrum consist of a narrow range in the electromagnetic spectrum which is between 400 nm and 700 nm. Prisms are used to separate white light waves into separate bands based on wavelength which we call the color of the rainbow. The flat band of colors is called the spectrum.

Let us now take a look at organization. Organizational components include dominance, proportion, balance, variety, harmony, repetition, movement, and selectivity. We are going to discuss each one of these components. Artists use dominant techniques to grab the attention of the viewer. In order to achieve dominance, an artist puts more emphasis on an image using either or a combination of color, line, shape, value, or a contrast in texture. Dominance can also be achieved by having an object in sharper focus than the rest of the surroundings. Proportion is another component when it comes to painting or drawing. There should be an agreement of the relative parts of the object and the intensity of the colors. Odd numbers rather than even numbers of visual components should be used for artwork to make it more interesting. Balance is also used to achieve equilibrium and harmony.

Usually in artwork, one side of an image is different from the other side. However, their can be a similarity and a sense of balance achieved. We call this asymmetrical balance. Figures in a painting do not have to be equal, but their can be similar shapes and movements. Harmony has to do with the combination of shapes. It is enhanced when variety is included. Repetition has to do with visual elements being repeated. Energy and movement is associated with repetition. Being selective for the work is also important. It is important to know what is and what is not essential to accomplish the purpose of the painting. We can see these are many factors to consider when an artist arranges a work of art.

When preparing for a work of art, it is important to have good composition. Arrangements of lines, shapes, colors, textures, and light-and-dark contrast can create an interesting work of art. Two- and three-dimensional artworks require elements to be the same to achieve a good composition. In two-dimensional artwork, color plays a role. In three-dimensional artwork, shapes plays a role. Good composition involves the elements of design such as line, texture, shape, color, and light-and-dark contrast. Any kinds of composition include elements of design to achieve a strong color of interest. This is where we attract the viewer's attention. This can be achieved by having a lighter interest against a darker area or the other way around. An intense color against a dull area or the other way around is something the artist can achieve. Another possibility is having a group of small objects against larger shapes or the other way around or a detailed area against a picture that is less complex. Contrasting the focal point against a composition is known as having a dominance or subordination effect.

Moving our eyes throughout the composition is also important. The rest of the composition should be arranged so the eye will move in a predetermined path. The movement can follow a direction such as going around, back and forth, up and down or diagonally. This can be carried out by repetition of a color in different amounts, lines, shapes, and textures repeating itself in various sizes. Parts of a painting should also be balanced. The focal point does not have to be put in the center of the picture. Normally, the focal point is moved off-center. Balance is important for good compositions. There are two types of balance: formal and informal. Formal balance involves the subject to be centered and other elements being placed on either side of the imaginary line to give a balanced presentation. Informal balances involve objects being off or away from the center. Creating interesting negative and positive space is important for good compositions. The artist should be aware of shapes and sizes of the subject and also pay attention to the negative space. A lot of negative space overrides the subject matter. Good composition should possess both unity and harmony if line, texture, shape, color, and light-and-dark contrast are done carefully. When we talk about unity, it does not mean "sameness". Size, shape, and color are important when it comes to unity.

The primary colors of light are yelllow, red, and blue. When all the colors of light are mixed, we get white light. When white light is absent, we get the color black. In order to see color, light enters the eye through the iris and focus on the retina. The retina contain rods for us to see at night and cones to detect color during the day. Light wavelength stimulates retina cones. This is where the brain translates into color. Most pigments do their best to absorb, transmit, or reflect wavelengths of light. Basic pigment colors include yellow, red, and blue. These are considered primary pigments because other colors cannot be mixed to produce these colors. We can observe a primary pigment's color because it depends on the color of light it reflects. Primary colors include yellow, red, and blue because there are no combinations of other colors that will produce other colors. The remaining colors are produced from the three primary colors. Secondary colors include orange, green, and violet. These can be made by mixing two primary colors in equal amounts. Intermediate colors are made by combining primary and secondary colors in equal amounts. Intermediate colors include red-orange, red-violet, blue-violet, blue-green, yellow-green, and yellow-orange. Neutral colors include white, gray, and black. Having a neutral color next to a color intensifies the color. We have to be careful when we use black and white. If it is not used properly, either of these colors can dominant a composition. The color black is seen when no electromagnetic radiation is seen with our eyes. White is seen when wavelengths from the electromagnetic spectrum is seen with our eyes.

Hue can be thought of as the response of our brains based on quality and quantity of wavelength. Each wavelength shows a different kind of hue. When there is no visible light, there is no color or hue. Artists like to use many kinds of colors to express an idea. Paintings are normally considered to be polychromatic. Value is the response of the brain to a quantity of wavelength that reaches the retina. When we add black pigments to a hue, light that is reflected from the hue is absorbed. The hue looks darker. When a white pigment gets added to a hue, a lot more light is reflected from the hue. The hue appears lighter. Intensity is the response of the brain to the purity of color. Pure colors are said to have high intensity. Laser light has high intensity for example. All light waves reaching the retina have the same wavelength. Hue also refers to color quality based on its name. We can change the color of a hue by mixing it with another hue. Value has to do with the lightness or darkness of a color. Light or darkness depend on how much white or black is added to the hue. If we want to make a hue lighter, we can add white to it. If we can make a hue darker, we can add black to it. Keep in mind adding white or black to a hue doesn't change the hue. It only lightens or darkens it. Intensity has to do with the strength of a color or its hue. Pure color is the strongest and most intense. The color's intensity is lessoned by adding gray to the color. This ends up dulling the color. The intensity of a color can be lessoned by adding a hue that is opposite based on the color wheel. The opposite color is called its complement. Again value can be thought of as lightness or darkness of a hue. We can see values in black-and-white works. Shades of gray can be used to define shapes and distinguish it from another. Amounts of light that strike the retina determine the lightness or darkness of an object. The amount of light reflected from a surface is determined by the light source, the kind of reflecting surface, the distance of the light source from the reflecting surface, and angle of reflected light. We can change the value of a hue by changing the amount of light reflected from the hue. We can alter the amount of reflected light by adding either black or white pigment to a hue.

An artist can change the value of a hue of pigment by adding either black or white pigment so the hue can become darker or lighter. Black pigments absorb wavelengths of visible light. When added to a pigment's hue, it reduces the number of light waves that reaches the retina. This makes the hue darker. White pigments reflect wavelengths of visible light waves. This increases the concentration of all wavelengths that reach the retina. The brain then interprets this as white light. The hue then appears lighter. We can also change the value of a hue of pigment by changing the concentration of hue particles in a pigment solution. Intensity has to do with color purity. Intensity can be changed by mixing color with a complementary color or by making the color look more gray. Again, intensity refers to the saturation of a color. Saturated colors are reached at maximum intensity. Artist lesson color intensity by adding gray to make it dull, or it can add a complementary color based on the color wheel. We can say the artist is changing the concentration of the particles that are colored in solution. In a saturated solution that is colored, color is at the maximum intensity. If for some reason, the solution is not saturated, the color becomes less intense since there are fewer colored particles. An artist can change the intensity of a hue by adding complementary color or by adding white or black pigment. Artist changes the hue's intensity by adding amounts of complementary color. When complementary color is added, wavelengths of light get reflected by the original hue then get absorbed by complementary color particles. Less light of the original hue is seen. Therefore, the color becomes less intense. In terms of a chemist's view, the intensity of a color can be determined by how much colored producing solute particles there are in a given volume. It makes sense the more particles that are colored, the more intense the color has to be. Let us also consider relativity. How we combine colors placed next to each other changes the way we perceive the colors. The complementary colors that are placed next to each other can make them more brilliant. A specific color can appear one way against a black background and a different way against a light background.

Paint

Objects are painted to prevent rusting and to make sure there is no deterioration. It also makes objects look attractive. Artist paint for attractiveness and emotional effect. During ancient times, paint pigments have been found in plants, animals, and minerals. Some of the main colors during the Paleolithic period were earth tones of yellow, red, brown, white, and black. These colors came from plants, animals, and minerals mixed with animal fat. Paint can be defined as a solution, suspension, or colloid made of a colored pigment and a binder. The binder dissolves the pigment and sticks to the surface. The resulting paint can either be transparent or opaque that depends on the binder. Mediums are used to dilute paint. Additives found in paint include glycerin and odorants. Salts containing lead, cobalt, or manganese can be dissolved in oil to speed up the drying process. Varnish could also be used to thin oil paints. Suspension agents could also be added to make sure paints do not harden in a metal tube. Antioxidants could also be added to make sure paints do not spoil.

Paints can be solutions. Recall that solutions are made up of a solute and a solvent. Pigment is considered to be the solute. The binder is considered to be the solvent. Paints are homogeneous mixtures. Paints can also be colloidal solutions. The particles do not come together and form larger particles. Colloidal particles have electrical charges of the same sign. Paints can also be mixtures. They are made up of pigment particles and binder particles. They are uniform mixtures, and they are also considered to be solutions.

Let us take a look at some specific examples of reactions involving art. Lead containing pigments can react with hydrogen sulfide in air to give a black precipitate lead sulfide. The reaction can be represented as follows

$$PbCO_3(s) + H_2S(g) \longrightarrow PbS(s) + H_2O(g) + CO_2(g)$$

However, in the industrial atmosphere the compound hydrogen sulfide gets oxidized to sulfur dioxide and dihydrogen sulfate. Lead-based pigments converted to sulfide-based pigments contribute to darkening over time. White lead and chrome yellow which contain lead pigments should not be mixed with ultramarine. Ultramarine is a complex mixture of sodium aluminum silicate and sulfide. Black lead sulfides are present when mixing of these dyes are used. Other chemical factors lead to dark pigmentation such as the following reaction

$$Cu_2O \longrightarrow CuO$$

Cuprous oxide, Cu2O, is a bright red pigment that gradually gets oxidized to cupric oxide, CuO, which is black. Exposure of light causes copper resinate which is green to become dark brown. Ferric ferrocyanide mixed with basic pigments such as carbonate white lead precipitates ferric hydroxide which gives a reddish tinge to the paint. The following reaction represents the net balanced equation

$$Fe_3+(aq) + 3OH-(aq) \longrightarrow Fe(OH)_3(s)$$

It is important to take care of each painting in order to maintain its beauty. Mediums is another factor artists need to take into account. Each artist has a specific medium that they like to choose whether it is watercolor, egg tempera paint, oil paint, or acrylic paint. When using transparent watercolor, colors are applied by putting thin washes. Colors can be built on top of another color. The white portion of the paper shines through beneath the paint. Adding water helps to lighten the colors. Transparent watercolor contains ingredients. These include the pigment, binder, gum arabic, and the sap of the acacia tree. Glycerin is put in transparent watercolor tubes to help the brush better. Ox gall is used for absorbency. A preservative is also added to make sure the paint lasts longer. Odorants such as oil of clove give the paint a nice smell.

Egg tempera paint involves having a powdered pigment bound with egg and water. Paints are applied layer after layer which gives a beautiful luminous quality. Oil paints give depth and also a beautiful luminous quality. It dries slowly, and it can be applied as a thick or thin coat. Acrylic paint can be used with a variety of media. It can be applied as a thick or thin coat. Tempera paint is water-based paint. It is considered to be water-soluble. But it is normally used for lower grade students.

Paints are made in a specific way. Pigments are made by grinding minerals that occur in nature. The finer the mineral is grounded, there is a better chance making a smooth homogeneous paint solution. Paint pigments can be formed by precipitation. Yellow lead chromate, blue copper carbonate, and white zinc hydroxide can be formed as follows

$$Pb(NO_3)_2(aq) \ + \ Na_2CrO_4(aq) \ \longrightarrow \ PbCrO_4(s) \ + \ 2NaNO_3(aq)$$

$$Cu(NO_3)_2(aq) \ + \ Na_2CO_3(aq) \ \longrightarrow \ CuCO_3(s) \ + \ 2NaNO_3(aq)$$

$$Zn(NO_3)_2(aq) \ + \ NaOH(aq) \ \longrightarrow \ Zn(OH)_2(s) \ + \ 2NaNO_3(aq)$$

Precipitating aqueous ions in solution help create colored pigments. Recall that precipitation occurs when charged ions combine based on solubility rules to form a solid substance in aqueous solution. The precipitate is used as a paint pigment. Precipitates can be separated by filtration. Other paint pigments include iron oxide which is red, copper oxide which is blue, and lead oxide which is yellow. Copper can be combined with acetic acid to make copper acetate which is green. Heating lead coils in acetic acid fumes with carbon dioxide in a moisture environment gives lead hydroxide and lead carbonate which is white.

Pigments can also be prepared by combining metals with oxygen to make metal oxides. These pigments that make up paints form mixtures. There are different types of mixtures such as a suspension, a true solution, or a colloid. Solutes in a suspension have particles less than 1,000 nm. Solutes in true solutions have particles less than 1 nm. Colloids have solute particles between 1 nm and 1,000 nm. Colloids can be described as a state of matter between a solution and a suspension. Colloidal particles do not settle out in solution, instead they scatter light. Pigments combine with a binder to form a colloid. A lot of paints are colloids, but they can also be chemical solutions. The best paints cover the surface in a uniform fashion and dry in a proper amount of time.

Clay

Clay is considered to be both plastic and pliable. It can be pushed, pulled, pounded, and pinched. Ligand forms of clay can be poured into mold. Clay is considered to be the powdery form of rock broken into fine particles. The formula for clay is $Al_2O_3*2SiO_2*2H_2O$. Kaolin is considered to be primary clay. It is formed by the weathering of feldspar. They are coarse in particle size and do not have mineral impurities. Secondary clay has fine particles and could be contaminated with iron, mica, quartz, and carbonate compounds. Examples of secondary clays are ball clays. They are also more plastic than kaolin clays. Terra cotta is clay that has open coarse grain structure which allows for fast drying. It is used to make large clay pores. Natural clay was formed from geological processes on the surface of Earth. Rocks can be broken down into small particles. Mountains form by these rocks which erosion tears them down. The resulting silt gets deposited in stratified layers. The process repeats which leaves clay deposits.

Let us look at different types of clays. Natural clay is soil beneath the top soil of the surface. Feldspar which consist of potassium oxide, aluminum oxide, and silicon dioxide breaks down through chemical reactions to form natural clay. Adding other substances can change both the color and texture of the clay. Natural clay can become plastic and act cohesive when it is moist. Platelets of silicon, oxygen, hydrogen, and aluminum atoms bond together in a arrangement that slides one another with water. Natural clay can get hard if it is exposed to heat. Baked clay pottery has a rough and porous surface. Giving clay a smooth and shiny surface is called glazing. Other types of clay include plasticize, low fire, and self-hardening clays. Plasticene clay has synthetic polymers which does not harden. However, it cannot be fired and glazed to be permanent. Low-fire clay comes in many colors. It can harden when baked, but it cannot be glazed. Self-hardening clay hardens in air and dries to a nice finish. Kilns can be used for heating or firing clay.

Ceramics

Ceramics has to do with molding, firing, or heating clay to high temperatures. It results in hard and permanent objects. Ceramics include items such as kitchen sinks, bathtubs, floor tiles, and walls. Natural-clay ceramic must be kneaded in order to remove pockets of air to make it smooth and homogeneous. Leather-hard clay is made when natural clay slowly dries to a particularly dry state. Greenware is clay that is hundred percent dry. Bisque ware is also clay heated in a kiln at high temperatures to make it dry, hard, and permanent. Glaze is glass that sticks to the surface of the clay. Ingredients in glaze include barium oxide, magnesium oxide, calcium oxide, strontium oxide, and zinc oxide which act as fluxes in glazes. The compound silicon dioxide helps increase the fusion point of glaze. Diboron trioxide is used to make low temperature glazes. Most oxides found in glaze are found in feldspar. The glaze pieces, the clay, and glaze oxides become one. Glazes can be shiny or dull. A dull finish is referred to as a matte finish. A shiny finish is referred to as a glass finish.

Glass

Glass is considered to be a noncrystalline substance that can either be transparent or translucent. It is also thought of as an elastic solid. Glass is made up of silicate anions with a tetrahedral shape.

They are arranged in a disorderly arrangement. It is a solid that has a definite shape and volume. However, over time it can act as a liquid and flow. Thick and thin portions of glass can be formed on old glass windows. Glass can be melted and then cooled to form elastic, transparent, or translucent which lacks a regular lattice structure. This is different from crystalline substances. Crystalline substances get melted and then cooled. Particles form into a regular lattice and repeats itself every time it is cooled. Glass has a wide range of temperature. Crystalline solids have a more fixed melting point. Note that crystalline solids are not glass. Glasses have an irregular shape, and it is considered to be amorphous.

Glass components determine the properties of glass. The compound silicon dioxide forms the body of a glaze. Sodium oxide, potassium oxide, and lead oxide are fluxes. They help lower the melting point of the compound silicon dioxide. The compound aluminum oxide hardens the melt. Calcium oxide helps the glaze to be hard and durable. Colors can also be formed on glazes. The compound cobalt oxide added to glaze and which contains magnesium oxide gives a purple color. Combining copper oxide and zinc oxide makes turquoise green. The compound diboron trioxide added to glaze which has some iron oxide can make the glaze milky or form a blue color. The ionic solid sodium carbonate can be added to glass which forms a more common kind of glass. Adding the compound potassium oxide to glass can make a very hard glass. Glass can be colored by adding transition metal elements. Adding copper(II) oxide to glass gives a blue-green color. Adding chromium(III) oxide to glass gives the color orange. Adding cobalt(II) oxide to glass gives the color blue.

Conclusion

We can see now that color has chemistry involved in it. It is the atoms and molecules that make up colored pigments. White light is composed of all the colors of the rainbow. It can be broken into different colors into a prism. Color is made up of different wavelengths and how we perceive it in light. How we perceive color and how we use it is up to the artist and the chemist. The primary colors are yellow, red, and blue. Secondary colors include orange, green, and violet. Mixing various colors together can produce a wide range of other colors that depends on both the artist and chemist. Intermediate colors include red-orange, red-violet, blue-violet, blue-green, yellow-green, and yellow-orange. Neutral colors include white, gray, and black. Having a neutral color next to a color intensifies the color. Balance and imbalance is another factor that is taken into account when color is created for the artist and chemist.   An artist can make a real life painting by mixing colors in different proportions based on hue, value, and tone. Similarly, the chemist can mix different colors of chemicals in various stoichiometric amounts to produce the desired end product from its reactants. A saturated solution that is colored has maximum intensity. The color becomes less intense since there are fewer colored particles. A change of intensity of a hue takes place by adding a complementary color or by adding white or black pigment. Wavelengths of light get reflected by the original hue then get absorbed by complementary color particles. Less light of the original hue is seen. The intensity of a color can be determined by how much colored producing solute particles there are in a given volume. Creativity and imagination are important for the creation of new substances.

Objects can be painted to make sure there is no rusting and there is no deterioration. Chemists make up the paint solutions, and artists use them to create attractiveness and emotional effects. Paint pigments are found in plants, animals, and minerals mixed with animal fats. Both pigment and binder make up the paint solution. The paint itself can be transparent or opaque which depends on the binder. Solutions are made up of solutes and solvents. The pigment is considered to be the solute, and the binder is considered to be the solvent. We can think of paints as homogeneous mixtures. Mediums such as watercolor, egg tempera paint, oil paint, or acrylic paint can be used. Precipitates can be formed in solution that create colored pigments. They can be separated out by filtration. Different types of mixtures include suspension, colloid, or a true solution. All three types of mixtures depend on the size of the particles.

Clay is a powdery form of rock that has been broken down into fine particles. Kaolin is primary clay. Secondary clay have fine particles probably contaminated with iron, mica, quartz, and carbonate compounds. Ball clays are secondary clays. Terra cotta is clay that has open coarse grain structure for fast drying. Natural clay is soil that is beneath the top soil of the surface. A process called glazing can give a smooth and shiny surface. Other kinds of clay include plasticene, low fire, and self-hardening clays. Ceramics has to do with molding, firing, or heating clay at high temperatures. Glass is considered to be a noncrystalline substance that is either transparent or translucent. This is different from crystalline substances which have an ordered arrangement. Glass has an irregular shape, and it is amorphous. Colors can be formed on glazes by adding specific oxides. Glass, too, can be colored by adding transition elements. More materials are expected to be made in the future.

Chapter 25

Cosmetic Chemistry

Cosmetico

The purpose of cosmetic chemistry is to introduce the fundamental principles of cosmetic formulation. It is the intent of this chapter to provide formulas and to have a general science background. Many women open their makeup bags and apply lipstick, mascara, and other kinds of make-up without knowing what they are putting on. Make-up consists of chemicals, and they need to be put on wisely for the best results. Understanding the chemistry of each of the cosmetics should help make a wise decision choosing the right product.

The purpose of chemicals in cosmetic products is to improve and refine the appearance of skin, lips, lashes, and other body and facial features. However, applying make-up is not that easy since some of it can cause serious adverse side effects. The first thing to do is to read the label. The United States Food and Drug Administration makes it a requirement for cosmetic manufacturers to put labels on the listing ingredients in descending order of weight. Ingredients that contain less than one percent of make-up consist of fragrance or colorants. These are listed after the other ingredients. Make-up users need to be careful to reduce the chances of developing rashes, eye infections, acne, or other health related problems involving cosmetics. Consumers should not waste money on cosmetics that are expensive than those cosmetics that are less expensive.

Primers

A primer is considered a cream or lotion that is applied before putting on another cosmetic. This helps improve coverage and makes the amount of time the cosmetic to last longer on the face. Examples of cosmetic primers include foundation primers, eyelid primers, lip primers, and mascara primers. Foundation primers can work like a moisturizer. It can absorb oil with salicylic acid or help create a less oily appearance. It helps make the foundation to be smooth. Some of these foundation primers contain antioxidants (A, C, and E). Some of these primers also contain grape seed extract and green tea extract. There are two types of foundation primers. These include water-based and silicon-based. They contain the ingredients cyclomethicone and dimethicone. Note that some of these primers don't contain preservatives, oil, or fragrance. Some of them can have sun protection factor and even improve skin tone. Foundation primers can also be mineral-based primers that contain mica and silica. Eyelid or eye shadow primer evens out the color of the lid and the area near the upper eye. It could reduce oiliness and add shimmer. These primers help smoothen the application of eye shadow and stop it from building up in eyelid creases. Eyelid primers are put on the eyelid and the lower eye area before eye shadow is applied. The color of the eye shadow is intensified, and it helps them from smearing by the reduction of oiliness of the lids. Eye shadow primers can also work for eyeliners and eye shadow bases. Other primers include mascara primer. Some of the mascara primers are colorless. It thickens and lengthens the lashes before mascara is applied. The lashes last longer when primers are used. Lip primers smoothen the lips and improve the appearance when lipstick is applied. It is recommended lip-gloss be used to exfoliate the lips before applying the lip primer. Color lasts longer and to make sure the lipstick does not smear and migrate into the lines around the lips.

The bottom line is foundation primers allow the foundation to stay on during hot weather. They help prevent the skin from absorbing talc and pigment from the foundation. It also prevents talc from drawing oil from the skin. To soothe and refreshen the skin, botanicals such as lavender, grape, kiwi, rose, jasmine, orange, and aloe extracts are added and used in the foundation process.

Lipstick

Lipstick consist of waxes, and emollients. Lipstick is used to put color on, to add texture, and to protect the female's lips. Waxes give the structure of solid lipstick. Waxes can be made from beeswax, ozokerite, and candelilla wax. Carnauba wax has a high melting point, and it also provides the strength of the lipstick. Oils and fats are also found in lipsticks, pigments, and oils. These include olive oil, mineral oil, cocoa butter, lanolin, and petrolatum. Most of the lipsticks in the United States contain pig fat or castor oil. These give a shiny look. The color of lipsticks comes from a variety of pigments and lake dyes. These include bromo acid, D & C Red No. 21, Calcium Lake (D & C Red 7 and D & C Red 34), and D & C Orange No. 17. Mixing colorless titanium dioxide and red shades gives pink lipsticks. Both organic and inorganic pigments are present.

Matte lipsticks have more filling agents such as silica, but they do not have a lot of emollients. Crème lipsticks have more waxes than oils. Sheer and long-lasting lipstick has a lot of oil. Long lasting lipstick has silicone oil. This seals the colors to the person's lips. Glossy lipstick has more oil and provides a shiny finish to the lips. Lipstick that looks shimmery has mica, silica, fish scales, and pearl particles.

Let us look at the chemistry of lipstick. The pigments in lipstick can be separated using thin layer chromatography. The mobile phase can vary depending on the kind of pigment. Lipsticks are considered to be soluble in toluene. Normally, toluene is used as the mobile phase. Once the separation process is complete, a chromatogram shows the different pigments that compose the lipstick. Lipstick can be made from grinding and heating specific ingredients. Waxes that are heated are added to the mixture. Oils and lanolin are also added. Then the hot mixture is poured onto a metal mold. The mixture is then cooled so the lipstick can harden. After the lipstick is hardened, it is heated in a flame for less than a second to give a shiny appearance.

Concealer

Concealers are considered to be color correctors. The purpose of concealers is to mask dark circles, age spots, and blemishes that are seen on the skin. It is thicker than foundation, and it has different pigments. It does this by blending around the skin. Concealers are applied after the primer and before the foundation on the face. Using concealers and foundations make the skin more even toned based on its color. Concealers have more pigments in them. It can be found in the liquid or solid form. Foundation is normally put on larger areas. Women can use concealers by itself or with foundations.

Concealers are available from the lightest and deepest shades. Women have a tendency to choose shades lighter than their skin tone to hide blemishes and dark circles under the eye. Some of the colors give the natural skin tone, while others are used to contrast with a blemish. Concealers that have yellow undertones hide dark circles. The colors green and blue counteracts red patches on the skin-like pimples, broken veins, or rosacea. Concealers that are purple make the complexions look brighter. Colored concealers are applied a little bit beneath a concealer or foundation to match the female's skin tone. Concealers that are skinned toned are sufficient at covering up imperfections.

Common concealer ingredients include talc, macadamia oil, and ground mica. Titanium dioxide and shea butter is also found in many concealers. Talc and titanium oxide could irritate the lungs if someone breathes into them. Some people are also allergic or could get skin irritation from macadamia oil, mica, and shea butter. Normally, other kinds of minerals and vitamins become mixed to make the final product. Talc gives a nice shimmer. Macadamia oil has a slight odor, but it is covered up with other ingredients. Ground mica reduces the appearance of wrinkles and other blemishes. Mica also protects the skin by blocking rays from the sun. Titanium dioxide protects us from the sun. It forms a barrier between the skin and harmful rays.

Foundation

Foundation is considered to be a skin colored cosmetic put on the face to make a uniform color to hide flaws and to change the skin tone. Both coverage and formulation is used in foundation, opacity of makeup, transparency, and contains a small amount of pigment. It does not hide discolorations on the skin. It minimizes contrast between discoloration and skin tone. Light coverage covers blotchiness, but it does not cover freckles. Medium coverage can cover freckles, discolorations, and blotchiness. Full coverage is the most opaque. It is used to cover birthmarks, hyper pigmentation, and sears. Formulation has to do with ingredients blended together. Oil and emollient formulations consist of oil and emollient. Sheer coverage is pigment added to it. The texture is both thick and dense. Oil-based shakers can be applied to the skin with a texture that is smooth. Liquid foundation works well around the eye. Alcohol-based formulations contain water and denatured alcohol as a base with pigment added to it. They don't clog pores and provides a sheerest coverage. Powder-based formulations uses powder mainly talc as one of the ingredients. Pigments, emollients, skin adhesion agents, and binding agents are added to the formula. Mineral makeup refers to foundation as loose powder. Common minerals include mica, bismuth oxychloride, titanium dioxide, or zinc oxide. Water-based formulations with emulsifiers formed a creamy liquid giving medium coverage.

Water-based cream has a creamy texture. Water-based oil-free products contain an emollient ester or fatty alcohol, and it has a mattifying agent called day. The oil-free mixtures are thick and heavy. Water-based transfer-resistant uses a polymer to give a matte finish. Silicon-based formulations contain silicone or a mixture of water and silicone. Normally, dimethicone and polysiloxane are used. Volatile silicones are also used. Silicone gives lubrication and viscosity to be put on the skin more evenly.

Face Powder

Face powder is powder put on the face. It sets a foundation ready for makeup to be applied. Two types of face powder include translucent sheer and pigmented powder. Some pigmented facial powders are worn alone that has no base foundation. Powder gives the face a more even appearance. Some powders also have sunscreen that reduces skin damage from sunlight and stress from the environment. Loose powder applied gives a uniform distribution. There are many colors of face powder and several types of it. Talc or baby powder is considered to be an absorbent and gives the tone of the skin. Face powder needs to be chosen carefully to match the tone of the skin.

Blush & Bronzer

Blush is also known as blusher or rouge. Women use blush to redden the cheeks to give a youthful appearance. It also emphasizes the cheekbones. Blush is made up of red talcum powder applied with a brush on the cheek. The coloring contains either the substance of safflor or carmine in ammonium hydroxide and rose water mixed with rose oil. Schnouda is also considered to be rouge. The mixture is colorless and contains Alloxan and cold cream. The purpose of bronze is to darken the complexion. You have to choose one based on skin tone. It should be two shades darker than your skin tone. It is best to use specialized bronzer brushes. Bronzer is added after the concealer and foundation has been put on. Put it on evenly to smoothen your complexion and also make a blank canvas for contouring. It should also be blended down the neck slightly.

You want to swirl the brush evenly inside the bronzer. Apply in light coats that are even to build up color. You just want to lightly coat the brush in the powder and put the excess on the lid of the container. Make sure you put on the bronzer on your face where sun's rays will hit the most. Use light strokes and start at the top of the forehead, the cheekbones, and down the jaw. If too much bronzer is put on, the bronzer can be blended down with a little bit more base powder.

Mascara

Mascara is used to enhance the eyes. It could darken, thicken, and lengthen the eyelashes. It can be in the form of a liquid, cake, or cream. They contain the basic components of pigments, oils, waxes, and preservatives. The pigment for black mascara is carbon black. Brown mascaras use iron oxides. Some of them contain the pigment ultramarine blue. Specific oils such as mineral oil, linseed oil, castor oil, eucalyptus oil, lanolin, turpentine oil, and sesame oil is used. Paraffin wax, carnauba wax, and beeswax are common waxes found in mascara. The effects of mascara depend on the ingredients. An effect could be if the mascara is water resistant or not. Water-resistant mascaras have nonpolar substances such as dodecane. Non water-resistant mascaras have ingredients that are water-soluble. Mascaras that lengthen or curl eyelashes contain nylon or rayon microfibers. Ceresin, gum tragecanth, and methylcellulose are added to mascara to act as stiffeners.

Proper disposal of mascara should be done after three months. If there is ever a bad odor of mascara, it should be disposed of properly. Mascara can grow bacteria so one should be careful. It is rare but possible that old mascara can cause eye infection or conjunctivitis, swollen eyelids, and stys.

Swollen eyelids and stys are considered to be allergic reactions. The allergic reactions are normally due to the methylparaben, aluminum powder, ceteareth-20, butylparaben, or benzyl alcohol.

Color

Let us take a look at the ingredient namely color. Make-up products give color and consistency. We think of red lips, shadowy eyelids, and puffy cheeks. A wide variety of colors are made on make-up racks. Many cosmetic chemists come up with dyes and pigments from many different compounds. Examples used to add color to our faces include coal tar, chromium oxide, aluminum powder, iron oxide, manganese, and mica flakes. Beet powder comes from plants. Pigments and dyes derived from animals include carmine which is a crimson pigment made from dried bodies of cochineal insect.

Common coloring agents are the coal tar colors. Coal tar is sticky, and it is a black liquid made by heating bituminous coal in large ovens where air is absent. The colors of coal tar are made from ring-shaped aromatic hydrocarbons that are purified from the tar itself. These coal tar colors are the only makeup ingredients the FDA uses for safe testing of final products. Unfortunately, many of these compounds have been slow to cause cancer when injected into experimental rodents. Some of the coal tar colors are banned while some of them are approved.

Once they pass safety tests, they are either given the designation FD & C (Food, Drugs, or Cosmetics) that means the color is safe for both internal and external uses. If the designation is D & C or Ext D & C then the compound is safe for external uses. Coloring agents such as D & C or Ext. D & C should not be applied where they are absorbed such as close to the eyes and on the lips which could cause blindness. Females who spread face foundation for their lips or eyelids before using lipstick or eye shadow should pay attention to the information provided on labels. Sometimes coal tar dyes can make or cause allergic or irritant reactions like rashes or inflammation. Yellow and red colors pose the problems.

Again, we can think of color coming from pigments and dyes. The compound titanium dioxide is considered to be a white pigment. Iron oxides can range in color from yellow, red, brown, and black. Mixing inorganic oxides and fillers makes face powders. Fillers are considered to be inert materials such as kaolin, talc, silica, and mica that are used to extend and develop colors. Mixing more ingredients such as oils and zinc stearate, and then pressing the mixture into pans make pressed powders such as eye shadows and blushers.

Eye shadows and lipsticks contain pigments called pearls. The pearls tend to sparkle and reflect light to produce different kinds of colors. To prepare them, one precipitates a thin layer of color on mica that has thin platelets. If you vary the thickness of the color deposited, the angle of light changes, and it is refracted through the composite which then gives different colors. Organic pigments are also used to add color to lipsticks and eye shadows. If the organic compound is precipitated on a substrate, we call them laker pigments. We use the term lake to refer to the precipitating of an organic salt on a metal substrate. These are called D & C (drug and cosmetic) and FD & C (food, drug, and cosmetic) colors. Some of the dyes are soluble, and some of them are insoluble. Dyes give tints for lotions, oils, and shampoos. Speaking of color, hair coloring products have the ability to mask or remove gray in the hair. Hair-coloring products react with protons in hair. These kind of products give a black-colored compound to cover up the gray in the hair. Hair-coloring products contain lead acetate, $Pb(O_2CH_3)_2$, which is soluble in aqueous solution. A reaction occurs between the Pb^{2+} aqueous cation and the sulfur atoms in cysteine and methionine that is also present in amino acids found in hair. The product of the reaction is lead(II) sulfide, PbS, which is insoluble.

<u>Bases</u>

Developing colors that seem attractive can be a challenge for chemists. Finding a way to make these colors stay on the face for many hours can also be difficult especially through times of perspiration, eating, or drinking. Make-up consists of another major ingredient class called bases. Almost every kind of makeup requires an oily mixture, which is considered a base to hold the colors together in a tube. This helps the colors stay on the face. The bases discussed have to do with manufacturing.

The kind of base that is supposed to be used will depend on where it is supposed to be applied. Lipsticks consist of half of its weight by a thick insoluble mixture of waxes and castor bean oil. This does not dissolve when a female licks her lips or even drinks water. The lipstick base has to match the chemical properties of the ingredients in oil with wax. The oil found in the lipstick makes it thick and sticky so that the color stays on the lips. The waxes are considered to be thixotropic which means the waxes become fluid when stirred. They become thixotropic so the lipstick keeps its shape and so it does not smear or melt when heated.

Lipsticks also consist of esters. These are slippery chemical compounds. They are made by reactions between alcohols and acids. Esters make the lipstick shine and make dry oil and wax mixture smooth on the lips. Mascara, likewise, consist of heavy bases like paraffin and carnauba wax. They keep lash-darkening pigments stick to the eyelashes by water and tears. They also thicken and separate the lashes. Eye shadow, blush, and some powdery products stick together by lighter bases such as mineral oil. Most of the bases used for face foundation consist of mineral oil and water. The mixture of water and mineral oil gives an emulsion where tiny drops of a liquid are suspended in the second liquid. To help them stay together, emulsifying agents such as sodium stearate are added to make a creamy mixture.

Other bases include isopropyl lanolate or wool alcohol, myristyl lactate, and octyl hyroxystearate. These nontoxic compounds are called fatty esters. What is nice about bases is they cause very few allergic reactions. Lanolin products come from sheep's wool. Beeswax more likely does so too. Oils and waxes in makeup can cause acne for teenagers and young women. These greasy compounds give whiteheads, blackheads, and pimples because they can clog skin pores. Females who are prone to acne should carefully choose cosmetics that are both water soluble and free of oil.

Bulking Agents

Bulking agents are important ingredients in products that require even coverage like face powder and eye shadow. Examples of bulking agents include talc or French chalk, which is a powder, made from the mineral magnesium silicate. Talc is used in makeup because it absorbs perspiration. It also has a smooth and slippery texture. This helps make cosmetics easier to apply. You should never inhale deeply when using products containing talc like face powders, eye shadows, and powder blush. If you keep inhaling talc, you can develop lung problems. Other bulking agents such as silk powder can be added to eye shadow. Nylon and silk fibers can also be added to mascara. Be careful when using these products because silk powder triggers severe allergic reactions.

Waxes

Waxes make up long chain esters that remain as solids at room temperature. Waxes are used in cosmetics. Specific waxes include beeswax, candelilla, carnauba, paraffin, and polyethylene. Waxes are found in lip balms and sticks. Waxes work by being structuring agents, and it makes the stick rigid enough to stand on its own. It also forms properties involving barriers. If you combine waxes with unique properties like brittleness, high shine, and flexibility, then the best cosmetic performance can be achieved. Waxes can also be combined with compatible oils to achieve softness. Turbidity and the separation of materials mixed together above their melting points determine compatibility. Waxes have been found to be useful for hand creams and mascara emulsions for thickening and waterproof properties.

Thickeners

Mixing wax with thin lotion, a thick cream is formed. A lot of the thickeners are considered to be polymers. Cellulose, for example, is considered to be a polymer of repeating D-glucose units. Cellulose tends to swell in hot water making a gel network. Carpool is a polyacrylic acid. It swells when it is neutralized. Bentone clays also swell when opened up through mechanical sheer. Carrageenan, pectin, and locust beam gum are also examples of cosmetic thickeners.

Active Ingredients

Active ingredients are materials that work within the skin or help to protect the skin. An example is fruit acids. Fruit acids are also called alpha hydroxyacids or AHAs. They penetrate the skin where they increase the making of collagen, elastin, and intracellular substances that help improve the appearance of the skin. Many cosmetic active ingredients lighten, tighten, and make the skin firm. Active ingredients can also be used to suppress perspiration such as aluminum chlorohydrate. Other active ingredients include salicylic acid and benzoyl peroxide because of their ability to fight off acne. Petrolatum and dimethicone are also active ingredients to help protect the skin.

Sunscreens

The purpose of sunscreens is to filter out most of the sun's burning rays. Many consumers are concerned about skin cancer and skin-damaging effects due to excessive exposure of the sun and harmful ultraviolet light. Compounds such as oxybenzone and dioxybenzone can also be used to carry this effect. One should be careful, however, because these compounds can cause allergic reactions. The point is that sunscreens protect the skin from ultraviolet radiation. Ultraviolet radiation with wavelengths 290 nm and 400 nm damage the skin. They have the ability to absorb or reflect wavelengths. The compound para(PABA) can be screened out. Sunscreen has the reflect damaging sun-protection factor or SPF. If you use sunscreen that has an SPF of 15 that means you will be able to be in the sun fifteen times longer than if left unprotected. Sunscreens can include octyl methoxycinnamate, octyl salcitate, titanium dioxide, and avobenzone. They are classified either as UVA or UVB sunscreens that depend on the wavelengths they absorb. Benzophenone-4 is considered to be a water-soluble UV filter. It used to protect the color of cosmetics.

Preservatives

Preservatives are used to prevent microbial contamination and rancidity. Parabens and ester of parabenzoic acid are examples of preservatives because how effective they are against gram-positive bacteria. The compound phenoxyethanol has the ability to protect cosmetic products against gram-negative skins. A mixture of preservatives is used to protect against different bacterial strains, yeasts, and molds. Vitamin E and BHT are considered antioxidants that prevent oxidation of ingredients that are sensitive. It also protects the skin from free-radical damage.

Additives added to make-up include fragrances and preservatives. But they can account for allergic and irritant reactions. Cosmetic chemists decide to add fragrances to make-up for pleasing scent. Fragrances also hide bad odor of some waxes, oils, and some makeup components. Products that are labeled fragrance-free have no fragrance. If they are labeled unscented, they have no noticeable scent but have enough fragrance additives to cover up the smells of other ingredients. Basically, the purpose of preservatives is to kill microbes. They are another major kind of additive found in makeup. Bacteria and microorganisms reproduce a lot in warm solutions. If we did not have these preservatives, mascara, foundation, and other makeup becomes a culture for harmful microbes.

Chemicals are added to kill microorganisms or stop their growth. Parabens are examples of preservatives in makeup. Compounds such as butylparaben, ethylparaben, and methylparaben are normally not allergic. But be careful about preservatives called quarternium-I5, formaldehyde, and sorbic acid. These preservatives cause many allergic reactions. Another class of chemical preservatives are called antioxidants. They are listed as product labels such as butylated hydroxyanisole (BHA) and butylated hydroxytoluene (BHT). These additives are added to prevent ingredients combining with oxygen. When oxygen is combined, we call this oxidation, and it can ruin the makeup's color and texture.

Emulsions

Most creams and lotions are emulsions. We can think of an emulsion as two fluids that are immiscible. One of the liquids is dispersed as droplets in another. Oil can be dispersed in water. Fat does not float to the top because of emulsifiers. Water can also be dispersed in the oil phase. The nondispersed liquid is considered to be the continuous phase.

Surfactants

A lot of the emulsifiers are considered surfactants or surface-active agents. They reduce the surface tension of water. The hydrophile-lipophile or HLB balance determines whether the emulsifier surface is active or not. HLB can be determined by the size of the hydrophilic portion and the size of the lipophilic portion. The polarity of the material is important for the HLB system. Polar materials are towards the top of the scale, and nonpolar materials are towards the bottom of the scale. Emulsifier's polar groups orient towards the aqueous phase. The nonpolar groups orient towards the oil phase for the formation of micelles. Structures give stability to hydrogen bonding and weak electrical forces.

Skin-care emulsifiers are divided into groups of two based on ionic charge. Those that dissociate into charged species are ionic. Those materials that do not dissociate are called nonionic. Ionic emulsifiers is classified based on charge. Anionic molecules have negative charges when it is solvated. Fatty acids that react with alkali metals form soaps. The formation of soap is considered to be saponification. If the molecule overall is negative, it can be classified as anionic. Emulsifiers that are positively charged are called cationic. Amphoteric compounds have both negative and positive charges.

Nonionic emulsifiers are better to use for skin-care emulsion due to safety. They are normally grouped together based on similar chemistries. Fatty acids that are present in fats and oils are grouped together based on the length of the carbon chain. Fatty acids are one of the main components of emulsifiers because of their miscibility in both natural and synthetic oils. Polyethylene glycol and ethylene glycol are considered to be PEG esters. The solubility of a PEG ester is based on the number of PEG molecules that has been reacted per molecule of acid. Increasing the number of polar PEG molecules per acid increases the water solubility and the HLB is increased.

Emollients

Most emollients are considered to be fats and oils which we call lipids. The best emollients have spreading properties, low toxicity, low skin irritation, and the ability to have oxidative stability. Double bonds in some molecules give poor oxidative stability. Saturated fats and oils have no double bonds. There are also unsaturated oils that have double bonds which can react with oxygen when it is heated. Oxidation processes give unpleasant colors and odors in lipids making them rancid and not usable. More likely we will find petroleum-based emollients in formulations since they don't contain double bonds or reactive functional groups. Cyclomethicone and dimethicone are added to increase emolliency.

Essential fatty acids or EFAs present in oils have the ability to replenish lipids that are within skin layers. Fatty alcohols are best used as emollients and emulsion stabilizers. The polar hydroxyl groups orients itself towards the aqueous phase while the fatty side-chains orients itself towards the oil phase. Fatty acids and esters of fatty alcohols are good to use as emollients since they stability. Polar hydroxyl groups of sterols and alcohols make the grease absorb and hold water. Our skin is made up primarily of water. A lot of oils and emollients are used to take care and protect it. It has low reactivity and good composition.

Moisturizers

Moisturizers and emollients differ in their ability of being soluble in water. We need moisture to take care of our healthy skin. Moisturizers that are mainly polar are hygroscopic. In other words, they hold onto water. Measuring transepidermal water loss or TEWL can assess how effective moisturizers are. By applying moisturizer to the skin, the level of moisture is recorded. Eventually, the level of moisture reduces to the tendency of the skin so moisture is released over time. Ingredients that have a high level of moisture in the upper layers of our skin reduce the rate in which water can be lost. Glycerin helps reduce TEWL. Sorbital and sugars are used to hydrate the skin. Aloe has a mixture of polysaccharides, carbohydrates, and minerals. Together, they make a good moisturizer.

Skin Types

Five skin types include normal skin, dry skin, oily skin, combination skin, and sensitive skin. Normal skin has a smooth surface due to oil and moisture content. It is not greasy, and it is not dry. There are also small and barely seen pores. The skin looks clear, and it does not develop spots and blemishes. Little and gentle treatment is needed for people with normal skin. Taking care of the face is always required. Dry skin is different from normal skin. Dry skin has a tendency to flake easily. It is susceptible to wrinkles and lines because it does not retain moisture. Not enough production of sebum by sebaceous glands is produced. Cold weather becomes even more of a problem. Dry skin becomes oily even more. Moisturizer needs to be put on during the day. Moisture-rich cream needs to be put on during the night. Do not overexfoliate since the skin can dry out even further. Exfoliants like sugar, rice bran, or mild acids are the best to use, but they should only be used once a week. You want to make sure you avoid irritation and dryness.

Oily skin can be moderately greasy. This has to do with sebum being over secreted. A lot of oil on the surface of the skin creates dust and dirt from our environment to stick to the oil. Oily skin usually gives rise to blackheads, whiteheads, ugly spots, and pimples. Oily skin has to be cleaned every day. This applies especially in hot or humid weather. It is best to use a moisturizer that is both oil-free and water-based and is also a noncomedogenic moisturizer. It is important to carry out exfoliation.

Over-exfoliation causes irritation and increases the production of oil. It is best to use exfoliants that have fruit acids. Fine-grained exfoliants could help clear blocked pores.

Combination skin is also common. It consists of both oily and dry skins in different areas. Oily parts are normally found towards a central panel called the T-zone that consists of the forehead, nose, and chin. Dry areas make up the cheeks and areas around the eyes and mouth. Different parts of the face should be treated based on the type of skin. Sensitive skin is more delicate. Sensitive skin is more susceptible to irritation, redness, burning, flaking, rashes, and lumpiness. Chemicals, dyes, and fragrances cause irritation. Others include soaps, spice oils, flower oils, spray tans, tanning lotions, shaving creams, temperature changes, excessive exfoliating, threading, waxing, and bleaching. Avoid fragrances, chemical dyes, and those cosmetics that cause skin irritation. Sensitive skin could either be dry, oily, normal, or a combination.

Applying a cosmetic cold cream on the skin, for example, gives a cooling sensation because of evaporation of the alcohol. Skin products include ethanol. Alcohol helps the ability of the components inside the cold cream to dissolve. Having alcohol in the mixture helps putting the cream on the skin, gives a perfume odor, and it also gives a cooling effect on the skin. Evaporation of an alcohol gives a cooling effect because evaporation is an endothermic process or in other words it requires heat. The skin gives the proper amount of heat that is needed to evaporate the alcohol. This leads to a cooling sensation. The alcohol with a hydroxyl group attached to carbon is considered to be a volatile species. Alcohols such as ethanol have low boiling points and high vapor pressures that help with evaporation and which gives a cooling sensation. A higher vapor pressure for a liquid at a specific temperature causes a higher tendency for the molecules to escape to the gas phase.

Face Care

There are six ways we can take care of our skin. The first one is cleansing. Using a cleanser is the first step. Cleansers can be put on wet skin on the face and neck. Make sure you avoid the eyes and lips. Cleansing the face once every twenty-four hours should be sufficient. If makeup has been worn to remove extra dirt or oil, then a mild cleanser should be used. If you have oily skin, then you should clean it at least twice per day. Cleansers that are water-based are the best to use. But if you have acne, it is better to use medicated cleansers that contain benzoyl peroxide or salicylic acid. Soap should be avoided for dry and sensitive skin. Cleansers that are oil-based are good at removing dirt and makeup. It is always good to cleanse the face before putting makeup on.

Masks are a second way to take care of our faces. Facemasks are put on the skin for a certain amount of time, and then the mask gets removed. It is put on a clean se face making sure to avoid the eyes and lips. Clay-based masks make use of kaolin clay to put oils and chemicals on the skin. It is left on until it is completely dry. The clay then dries absorbing a lot of oil and dirt from the skin and helps clear blocked pores. Clay based masks should only be put on oily skin. Peel masks contain different types of acids or exfoliating agents that exfoliate the skin and other ingredients to hydrate the skin. These masks are left on to dry and then they peel off. If you have dry skin, then you should not use it. Sheet masks, however, are different. They contain a thin cotton or fiber sheet where there are holes cut out for the eyes and lips. Serums and skin treatments are then buried in a thin layer. Sheets can also be soaked for treatment.

Exfoliants is a third way to cleanse the face. They slough off both dry and dead skin. Acids or other chemicals is used to loosen old skin cells. Abrasive substances can also be used to scrub them off. The process of exfoliation evens out rough skin. This helps improve circulation to the skin, clear up blocked pores, and head scars. Exfoliants should be put on wet cleansed skin making sure to avoid the eye area. Abrasive exfoliants or scrubs can be used to rub into the skin by circular motion for around thirty seconds or more. If you have spots with severe flaking, dry skin should be exfoliated in those areas. This should be done only once per week. Some oily skins can tolerate twice weekly use of exfoliation. If you have sore, dry, irritated skin, or lots of dryness or oiliness, then this is caused by over-exfoliation.

Glycolic acid, lactic acid, salicylic acid, malic acid, acetic acid, and citric acid are considered to be chemical exfoliants. They could be in the form of liquids or gels. Some of them might contain abrasives to remove old skin cells. Microfiber brushes can be used to exfoliate the skin. You have to rub them on your face in circular motion. Creams, lotions, or gels might contain acid to help loosen dead skin cells. Abrasives such as beads or rice bran can be used to scrub dead cells off of the skin.

The fourth way we can take care of our face is through toning. Toners can be used after cleaning the skin and remove any traces of cleanser or makeup in order to maintain the skin's natural pH. Cotton pads can be wiped over the skin. It can also be sprayed on the skin from a spray bottle. Toners normally contain alcohol, water, and herbal extracts of other kinds of chemicals. Toners that have alcohol should be used for oily skins. If you have dry or normal skin, toners that are alcohol-free should be used.

Moisturizing is a fifth way to clean the face. Moisturizers are considered to be creams or lotions that hydrate the skin and help retain moisture. Some of them have oils, herbal extracts, or chemicals to help with controlling the oil or reducing irritation. NightGels, creams, and lotions are abrasive exfoliants. It can be applied to creams and, they are better hydrating than dry creams. Some moisturizers are tinted and contain a small amount of foundation. This provides light coverage for minor blemishes and to even out skin tones. Avoid putting moisturizers near the lips and around the eyes.

Regardless of the type of skin, it needs to be moisturized. Using a moisturizer reduces flaking and dryness. It could also help prevent wrinkles from forming. If you have dry skin, it is best to use oil-based moisturizers with ingredients so the skin can retain moisture and protect it from dryness, heat, or cold. If you have normal skin, you have many moisturizers to choose from. Light lotions or gels are the best ones to use. Moisturizers that are water-based should be put on oily skin. Moisturizers that are medicated and that also have tea extracts or fruit enzymes control oil production and could also treat acne. Our skin around the eyes is thin and sensitive. Eye creams are considered to be light lotions or gels. Some contain ingredients like caffeine or Vitamin K are used in order to reduce puffiness and dark circles seen under the eye.

The sixth and last way to protect our face is to avoid the sun as much as possible. The hot rays of the sun can cause a lot of damage to the skin including sunburns and skin cancer. The sun also exposes us to UVA and UVB radiation that causes uneven skin tone and dries out the skin. This reduces elasticity and encourages the formation of wrinkles. We should use sunscreen to protect ourselves from the skin. It also won't hurt to check the daily newspaper for the ultraviolet ratings index for the day. It gives a scale from 1 to 10 where 1 is the lowest amount and 10 is the highest amount of exposure of ultraviolet rays from the sun for the day.

We should put sunscreen about twenty minutes before being exposed to the sun. It should then be put on every four hours. Any skin that will be exposed to sunlight, sunscreen should be put on. About a tablespoon should be put to each limb, face, chest, and back for proper care. Sunscreens can be in the form of creams, gels, or lotions. The SPF number indicates how effective it is to protect the skin from the radiation of the sun. If you have oily skin, it is best to choose non-comodegenic sunscreens. If you have dry skin, it is best to choose sunscreens with moisturizers that help keep the skin hydrated. If you have sensitive skin, choose hypoallergenic sunscreen that is not scented. It might be a good idea to put it on a small spot to check to see if it does not irritate the skin.

It is important to get into the habit by carefully reading labels of ingredients and descriptions. Prone skin is a challenge facing several people. The best thing to do is to pick products labeled noncomedogenic. This means that they don't promote whitehead or blackhead pimples. You should also choose products that are nonacnegenic. This means they don't form any kind of pimple. It is also best to select products that are labeled oilfree. But do check the product if it contains greasy substances like lanolin, petrolatum, or emollient esters. Foundations do their best to prevent acne breakouts with ingredients such as salicylic acid. Salicylic acid fights microbes. Benzoyl peroxide acts as a drying agent. These ingredients dry the skin. But keep in mind a large number of people are allergic to benzoyl peroxide.

Dry skin can also be a problem. Dry skin becomes a problem for older women. Ingredients you should avoid include alcohols such as ethanol, methanol, and other types of alcohols. These dry the skin. It is a good idea not to use foundations or cover sticks that have powder in them. Some products have good drying agents. Cetearyl and stearyl are considered to be fatty alcohols. They help moisturize dry skin. Other good quality ingredients include lactic acid, glycolic acid, and urea. These nonoily compounds are called humectants, and they do not add moisture to the skin. Humectants cover the skin with a protective film that prevents water in the skin from evaporating. Sensitive skin and allergic reactions affect people of all ages. If the product is labeled "hypoallergenic" then they are less likely to cause allergies than regular products. Products that are labeled hypoallergenic normally do not contain fragrances, preservatives, and other causes of irritation and allergic reactions. Keep in mind these products are not 100% allergy proof. Some of the ingredients can cause allergic reactions.

Although products labeled allergy tested, sensitivity tested, and dermatologist tested are on makeup labels, they do not tell us exactly whether the product passed the test or not. If you are prone to allergies or have sensitive skin, it might be a good idea to test the new makeup on a different part of the body such as on your arm for a few weeks before applying the makeup on your face. If you wear contact lens you should be careful using mascara, eye shadow, and other eye makeup. Avoid mascara that has silk or nylon fibers and water-based mascara that falls into the eye. Frosted eye shadow is also not a good idea because the iridescent particles can fall into your eye and stick to the contact lens. It is possible this can scratch the eye's own lens which is in the cornea. You also need to reduce bacterial infection and contaminating particles. Be sure to wash your hands and put in your contacts before applying makeup. There are three basic safety warnings we should consider. The first safety warning is never put on mascara while in the car or when the vehicle is moving. A common injury when putting up makeup is scratching the eye using a mascara wand. The scratch itself is not the health hazard. It is the sight infection that can be threatening if the scratch is not treated right away. The second safety warning to consider is to never put makeup on that is spoiled. If the product does not look consistent, has a foul odor, or if there is a change in color it is possible it could have microbes. These can cause infections. Don't add water or saliva to the makeup that is dried. This will only add more contamination with microbes, and it can cause microbes with the liquid to be harmful. When you are not using makeup, it should be tightly closed and away from sunlight. Sunlight destroys preservatives. The third safety warning to consider is to never share makeup. Sharing makeup passes microbes back and forth. This is also a problem with sharing of testers at cosmetic counters. Counter makeup samples can be contaminated with bacteria, mold, and/or microbes. It is a good idea to test lipstick on your hand. Be sure to ask for a new cotton swab to test other makeup products.

Conclusion

It is important to understand emulsion chemistry and skin physiology when making personal care products. A lot of work has to be done to take care of us. Choosing the right product for our faces is crucial if we are going to achieve the best results of our facial appearance. A lot of the cosmetics discussed including primers, lipstick, concealers, foundation, face powder, blush, bronzer, and mascara needs to be carefully applied on the woman's face. Understanding the chemistry of each of these cosmetics can help make important decisions. It is good know something about the color, the bases used, bulking agents, waxes, thickeners, active ingredients, emulsions, moisturizers, sunscreens, preservatives, surfactants, and emollients. These determine the outcome of the product that is being used for the female. Taking care of the skin is just as important when choosing the appropriate cosmetic. Different skin types and the right use of facial care can help us look younger, natural, and appear nice in public. Despite all of the benefits looking nice using cosmetics, it is also important to consider safety features. Understanding healthy uses of cosmetics versus toxicity can keep us from making mistakes by choosing the wrong product. I hope you found cosmetic chemistry to be a fascinating subject, and I invite the reader to explore list of brands of cosmetics.

References

Zumdahl, Steven and Susan A. Zumdahl. *Chemistry*. Sixth Edition. Boston: Houghton Mifflin Company, 2003. Print.

Chang, Williams. *Chemistry*. Seventh Edition. New York: McGraw-Hill Higher Education, 2002. Print.

Greenberg, Barbara R. and Dianne Patterson. *Art in Chemistry Chemistry in Art*. Second Edition. Westport: Libraries Unlimited/Teacher Ideas Press, 2008. Print.

McMurry, John. *Organic Chemistry*. Third Edition. Belmont: Brooks/Cole Publishing Company, 1992. Print.

Timberlake, Karen C. *Chemistry An Introduction to General, Organic, and Biological Chemistry*. Ninth Edition. San Francisco: Benjamin Cummings, 2006. Print.

Karukstis, Kerry K. and Van Hecke, Gerald R. *Chemistry Connections: The Chemical Basis of Everyday Phenomena*. Second Edition. San Diego: Academic Press, 2003. Print.

Mittler, Gene and Rosalind Ragans. *Exploring Art*. Woodland Hills: The McGraw-Hill Companies, 2007. Print.

Stertka, Albert. *A Guide to the Elements*. New York: Oxford University Press Inc., 2002. Print.

Gray, Theodore. *The Elements A Visual Exploration of Every Known Atom in the Universe*. New York: Black Dog & Leventhal Publishers, Inc., 2009. Print.

Atkinson, Roger. Chemistry of the Clean and Pollluted Atmosphere (Reader)

Zieman, Paul. Chemistry of Natural Waters (Reader)

Significant Figures
http://en.wikipedia.org/wiki/Significant_figures

How Stuff Works, The Chemistry of Cosmetics
http://science.howstuffworks.com/the- chemistry-of-cosmetics-info15.htm

Cosmetics
http://en.wikipedia.org/wiki/Cosmetics

Primers
http://en.wikipedia.org/wiki/Primer (cosmetics)

Foundation
http://en.wikipedia.org/wiki/Foundation_(cosmetics)

Face Powder
http://www.en.wikipedia.org/wiki/Facepowder

Blush
http://en.wikipedia.org/wiki/Rouge_(cosmetics)

Bronzer
http://www.wikihow.com/Apply-Bronzer

Mascara
http://en.wikipedia.org/wiki/Mascara

Natural Health Techniques, Best Medical Intuitive Online
http://naturalhealthtechniques.com/list-ofcosmetics.htm

Cosmetic Chemistry
http://www.chemistryexplained.com/CoDi/Cosmetic-Chemistry.html#b

Printed in the United States
By Bookmasters